AUTOMOTIVE DIAGNOSTIC SYSTEMS

Keith McCord

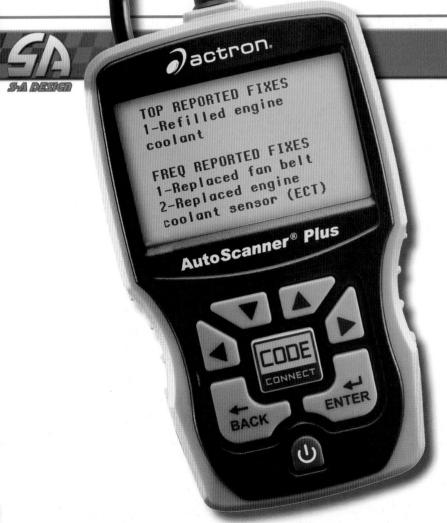

CarTech®

CarTech®

CarTech®, Inc.
39966 Grand Avenue
North Branch, MN 55056
Phone: 651-277-1200 or 800-551-4754
Fax: 651-277-1203
www.cartechbooks.com

© 2011 by Keith McCord

All rights reserved. No part of this publication may be reproduced or utilized in any form or by any means, electronic or mechanical, including photocopying, recording, or by any information storage and retrieval system, without prior permission from the Publisher. All text, photographs, and artwork are the property of the Author unless otherwise noted or credited.

The information in this work is true and complete to the best of our knowledge. However, all information is presented without any guarantee on the part of the Author or Publisher, who also disclaim any liability incurred in connection with the use of the information and any implied warranties of merchantability or fitness for a particular purpose. Readers are responsible for taking suitable and appropriate safety measures when performing any of the operations or activities described in this work.

All trademarks, trade names, model names and numbers, and other product designations referred to herein are the property of their respective owners and are used solely for identification purposes. This work is a publication of CarTech, Inc., and has not been licensed, approved, sponsored, or endorsed by any other person or entity. The publisher is not associated with any product, service, or vendor mentioned in this book, and does not endorse the products or services of any vendor mentioned in this book.

Edit by Paul Johnson
Layout by Monica Seiberlich

ISBN 978-1-934709-06-1
Item No. SA174

Library of Congress Cataloging-in-Publication Data

McCord, Keith.
  Automotive diagnostic systems : understanding OBD-I & OBD-II / by Keith McCord.
    p. cm.
  ISBN 978-1-934709-06-1
  1. Automobiles--Motors--Computer control systems. 2. Automobiles--Motors--Maintenance and repair. 3. Automobiles--Defects--Code numbers. I. Title.

TL214.C64M384 2011
629.28'72--dc22

2011003319

Printed in China
10 9 8 7 6 5 4 3 2

Back Cover Photos

Top Left:
The DLC connector commonly found on General Motors vehicles uses a variable-pulse-width communication protocol.

Top Right:
In this example, the data that the sender has sent is not equal to the data that was received. Unfortunately, the parity bit does not flag this error because an even number of bits are incorrect. This is the downfall of the parity-checking scheme because it can only detect when an odd number of bits are corrupt.

Middle Left:
Freeze-frame data shows the exact conditions when diagnostic trouble code P0301 (cylinder #1 misfire) was set.

Middle Right:
This is a partial listing of the data reported under the freeze-frame data menu. Typically, this information traverses several screens, so it is important to either print out the freeze-frame data if the scan tool supports it, or use the freeze-frame datasheet included in the appendix of this book.

Bottom Left:
A scan tool shows both bank #1 as reading lean after setting a P0171. Bank #2 is shown within specifications being 2 percent rich.

Bottom Right:
The data from the sender to the receiver is verified as the parity bit agrees between the two.

OVERSEAS DISTRIBUTION BY:

PGUK
63 Hatton Garden
London EC1N 8LE, England
Phone: 020 7061 1980 • Fax: 020 7242 3725

Renniks Publications Ltd.
3/37-39 Green Street
Banksmeadow, NSW 2109, Australia
Phone: 2 9695 7055 • Fax: 2 9695 7355

CONTENTS

Introduction 4

Chapter 1: Introduction to On-Board Diagnostics 6
- Closed-Loop Feedback Systems 6
- Pre-OBD Emissions Requirements 7
- Automotive On-Board Diagnostics 7
- Proprietary OBD: 1980–1987 8
- OBD-I 9
- OBD-I.5 12
- OBD-II 13
- Has OBD Made a Difference? 13

Chapter 2: OBD-II Standardization .. 14
- The Power of the Microchip 14
- Evolution of Automotive Networks 15
- Standardized ALDL Connector 16
- Standardized Scan Tool Data 16
- Diagnostic Trouble Codes 17
- The MIL 17
- Frame-to-Frame Data 18
- Real-Time Data 18
- What Information Does OBD-II Provide? 19
- SAE Standards versus ISO Standards 19

Chapter 3: The OBD-II Data Interface 20
- Data Link Connector 20
- Determining if the Vehicle is OBD-II 21
- OBD-II Data Protocols 22
- Serial Communications Protocols 22
- Troubleshooting Common DLC Connection Problems 24

Chapter 4: Scan Tool Interfaces 29
- Generic OBD-II Scan Tool 29
- Manufacturer-Specific OBD-II Scan Tool 29
- Reading Scan Data 30
- OBD-II Trouble Code Reader 30
- Entry-Level OBD-II Scan Tool 32
- Professional-Level OBD-II Scan Tool 33
- Personality Keys and Adapters 34
- Manufacturer-Specific OBD-II Scan Tool .. 34
- Personal-Computer-Based OBD-II Scan Tool 34

Chapter 5: OBD-II Modes 36
- Mode $01 – Request Data by Specific PID 37
- Mode $02 – Request Freeze-Frame Data by Specific PID 40
- Mode $03 – Request Set Diagnostic Trouble Codes 40
- Mode $04 – Clear Stored Diagnostic Trouble Codes and Reset MIL 41
- Mode $05 – Oxygen Sensor Test Results .. 41
- Mode $06 – Advanced Diagnostic Mode .. 41
- Mode $07 – Request On-Board Monitor Test Results 42
- Mode $08 – Control Operations of On-Board Systems 42
- Mode $09 – Vehicle Information 42

Chapter 6: Diagnostic Trouble Codes . 44
- OBD-II Drive Cycle 44
- Anatomy of a Diagnostic Trouble Code . 46
- Pending Diagnostic Trouble Codes 47
- Diagnostic Trouble Code Types 47
- Current and Historical Diagnostic Trouble Codes 48
- Plan Your Work and Work Your Plan 48

Chapter 7: Freeze-Frame Data 50
- Freeze-Frame Data Reports 50
- Freeze-Frame Data Summary 50
- Historical Freeze-Frame Data 51
- Breaking Down Freeze-Frame Data 53
- Using the Freeze-Frame Data Example .. 54

Chapter 8: Emissions Tests and Monitors 55
- Emissions Tests 55
- System Monitors 55
- Misfire Monitor 56
- Evaporative System Monitor 57
- Heated Catalyst Monitor and Catalyst Efficiency Monitor 57
- Secondary Air System Monitor 58
- Fuel System Monitor 58
- Oxygen Sensor Monitor and Heated Oxygen Sensor Monitor 58
- EGR System Monitor 59
- Comprehensive Component Monitor .. 60

Chapter 9: Four-Stroke Engine Cycle . 61
- Intake Cycle 61
- Compression Cycle 62
- Combustion Cycle 62
- Exhaust Cycle 63
- Otto Cycle Pressure versus Volume 63

Chapter 10: OBD-II and the Otto Engine Model 65
- Pressure and Vacuum 65
- Supercharging/Turbocharging versus Normally Aspirated 66
- Crank, Camshaft and Valves 67
- Static versus Dynamic Compression Ratios 67

Chapter 11: Controlling Fuel Systems .. 69
- Closed-Loop is the Key 69
- Fuel Combustion and Thermal Efficiency . 71
- Volumetric Efficiency 74
- Airflow Volume 75
- What Can Go Wrong? 75
- Evolution of the ECM 76

Chapter 12: Dynamic Fuel Correction . 77
- Fuel Requirements 77
- In a Perfect World 79
- Fuel-Trim Adjustments 80
- Real-Time Fuel-Trim Adjustments 80
- Historical Fuel-Trim Adjustments 82
- DTCs Related to Fuel Trims 83
- P0171 and P0174 DTC Fuel-Trim System Lean 84
- P0172 and P0175 DTC Fuel-Trim System Rich 84

Chapter 13: Engine Ignition Controls 85
- What is Ignition Timing? 85
- Generating High Voltage 86
- Distributors 87
- Knowing When to Fire 88
- Controlling the Timing 90
- Sensors that Affect Timing 90
- Detonation and Pre-Ignition 91
- The ECM and Detonation 94
- What Causes Detonation and Pre-Ignition? 94

Chapter 14: Misfires 97
- Misfire Types 97
- OBD-II Misfire Detection 97
- P030x Misfire DTC 98
- Diagnosing a Misfire 98
- Frame-to-Frame Data 99
- Blinking MIL 99
- False Misfire Code 99

Chapter 15: Sensors 100
- Intake Air Temperature Sensor (IAT) ... 100
- Engine Coolant Temperature Sensor (ECT) 101
- Throttle Position Sensor (TPS) 102
- Camshaft Position Sensor (CMP) 104
- Crankshaft Position Sensor (CKP) 105
- Manifold Absolute Pressure (MAP) Sensor 105
- Mass Airflow (MAF) Sensor 108

Chapter 16: Oxygen Sensors 110
- Oxygen Sensors Are Consumable 110
- How an Oxygen Sensor Works 110
- Diagnosing an Oxygen Sensor 112
- Oxygen Sensor Codes 113
- What Causes Oxygen Sensor Failure? ... 114
- Replacing an Oxygen Sensor 115

Glossary 116

Appendix A: Using a Volt Ohm Meter 117

Appendix B: Generic OBD-II DTC Codes 121

Appendix C: Manufacturer-Specific OBD-II DTC Codes 139

INTRODUCTION

In my youth we cultivated our automotive-repair passions early on and, most of the time, out of necessity. Back then, we all drove 1960s and 1970s cars that were not exactly known for their reliability, so we had to learn to diagnose and repair by the clunk-and-bang method just to be able to travel from point A to point B. Our tools consisted of no more than a timing light, a vacuum gauge, and, if you were lucky, a fancy analog dwell meter for setting your points. Along with our toolboxes full of tools, some of which never left the trunk of the vehicle for those on-the-road breakdowns, we learned all about the intricacies of the automobile.

As an engineer, I've watched the development of and worked with a variety of automobile control systems. I started with the simplest systems in the 1980s, which consisted of a couple of sensors, a blinking trouble light, and 4 kilobytes of software. I have seen the development of the modern systems, which feature self-diagnosing and multiple-redundancy vehicle-control networks with software modules in the 4-megabyte range. The speed and calculation capabilities of the modern automotive control module rival those of supercomputers from 10 years ago. Moore's Law says that technology in the computing world doubles capabilities every 2 years, and nowhere is this more evident than in the automotive world. Emissions requirements, safety requirements, fuel economy requirements, and creature comforts demanded by consumers continue to push the complexity of the automobile to new levels every year.

The technological advances are great for motor vehicle consumers, but for the automotive enthusiast, a simple timing light, vacuum gauge, and dwell meter are no longer sufficient to diagnose and repair the modern automobile. Luckily, the advent of the computer controls within the automotive industry has also given birth to excellent on-board diagnostic tools, also known as OBD. Also, the Internet has become an excellent source for information and discussions about automotive repair.

Unfortunately, for every good piece of information on the Internet there are probably 10 pieces of misinformation, so you still need to be skeptical and verify any information you find. Furthermore, many auto parts stores now read diagnostic trouble codes (DTC) for customers free of charge and even print out the diagnostic information for the vehicles. Regardless of their level of skill, and for a minimal investment, an enthusiast can purchase an OBD scanner and communicate with their vehicles to determine how well they are running and to help diagnose any issues that occur.

Everyone has seen it before, the dreaded Service Engine Soon light glowing menacingly on the dashboard. Often, the owner takes the vehicle to someone who scans the diagnostic trouble code and informs the owner on what the DTCs are and what parts need to be replaced. The parts are replaced and, within a week, the same Service Engine Soon light illuminates with the same trouble code and the cycle repeats, except now more parts are replaced. It can quickly become a reactionary approach, randomly replacing parts as diagnostic trouble codes appear, with no real path to predictive diagnosis. This methodology of diagnostics can rapidly become a money pit and possibly a safety issue for the owner.

The diagnostic issue isn't so much the complexity of the entire control system. The real issue is the ability to understand automotive problems and take a pragmatic approach toward solving them through systematic diagnostic procedures. To implement this

INTRODUCTION

approach, you need to understand how the sensors interrelate in the automobile, and how they feed data to the engine control module (ECM). It is important to know what the ECM does with this data and how it uses it to send informed instructions to operating systems. Understanding a system's functions is the first step in successfully diagnosing problems.

Automotive problems and diagnoses have morphed and intertwined during the recent decades, for the consumer and enthusiast. What remains the same is the often urgent need to figure out why the vehicle isn't running right, or to see why that darn engine light is lit on the dashboard. As my professor at the University of Missouri told me years ago, college prepares an engineer to gather data, diagnose issues, and solve problems—not so much to memorize formulas, equations, and such, but because life is one big word problem. And this is why I wrote this book.

My goal here is to pass on a basic understanding of how the control systems of a modern automobile interrelate and function as a cohesive group. By explaining and interpreting what is happening internally to the controls, and how to obtain information from them, this book offers a basic road map to help diagnose automotive problems. Unfortunately, there isn't any cookie-cutter process that says if diagnostic trouble code XYZ is found, replace parts one, two, and three. The automobile is far too complex to be reduced into a template like that. A glance at a dealership's service manual shows you that diagnosing an issue can take you to a variety of places within the control system, before possibly finding the troublesome part or system. The automotive system is basically one large closed-loop, feedback system. Something happens, something sees it happen, and something else reacts to it. It is really that simple.

This book is divided into three sections. First, I cover the on-board diagnostic system—what it consists of, how it works, and what information it can provide. Next, I cover the main functions of the engine—how it works based on the Otto engine model, how ignition is handled, and how fueling is determined. Finally, I look at the intricacies of the main systems and components that control a majority of the performance of the automobile.

It is impossible to cover every single sensor and every single operation that occurs in the wide range of vehicles on the market. I cover many of the common issues that enthusiasts encounter and can fix themselves. So, sit back, relax, read, and maybe re-read the chapters.

Most importantly, let this book help you learn basic fundamentals, so that you can establish a diagnostic path and better understand what is going on inside the "black box" that lives in every automobile.

CHAPTER 1

INTRODUCTION TO ON-BOARD DIAGNOSTICS

On-board diagnostics (OBD) are built into myriad systems used every day, including your personal computer, appliances in your home, and systems in your automobile. In short, on-board diagnostics are measurements and checks built into a vehicle's engine and systems, which ensure that the system is operating as designed or delivering what it is designed to do. If something is outside the prescribed operating conditions, an alert can be sent out, or other actions can be taken to bring the system back into balance. On-board diagnostic systems are based on a control engineering principle called closed-loop feedback systems.

Closed-Loop Feedback Systems

At the heart of any on-board diagnostic system is the concept of a feedback (closed-loop) control system.

In my simplified example, shown in Figure 1.1, a desired room temperature is set on the thermostat. The thermostat contains a thermometer, which measures the current air temperature. This measured temperature is compared against the commanded temperature. If the measured air temperature is lower than the desired room temperature, then the thermostat commands the furnace to turn on. The air temperature is continually measured until such a time that the air temperature is equal to or exceeds the commanded temperature. After that is established, the thermostat then turns the furnace off and continues the entire process of measuring the air temperature against the desired temperature.

Along the same theme, an example of a non-feedback control system is an old-fashioned iron stove. No thermostat controls existed on iron stoves used in the 1800s and early 1900s. If the temperature in the room got too hot, a person had to open the window to cool it down. If the temperature got too cold, the iron stove would not start on its own to bring the temperature back up. There is no intelligence in the old iron stove's performance. The key to a closed-loop system is: Action leads to measurement leads to correction of action.

On the modern computer-controlled vehicle, the main closed-loop feedback system is the fuel control system. Based on emissions, fuel economy, and performance standards, the closed-loop feedback system is constantly monitoring the real-time conditions of the engine and adjusting the fueling based on feedback from a variety of sensors. This closed-loop system has allowed modern-day

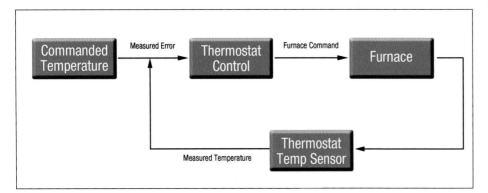

Figure 1.1. A simple example of a closed-loop feedback system is the thermostat that controls the home furnace, which in turn controls the temperature of a home. The feedback system ensures that the furnace of the area achieves its desired target temperature.

INTRODUCTION TO ON-BOARD DIAGNOSTICS

automobiles to eclipse the performance, reliability, emissions, and fuel economy of any of the pre-computer-controlled automobiles.

Pre-OBD Emissions Requirements

In 1967, California created the California Air Resources Board (CARB) in order to help establish and enforce emissions standards in that state. In 1970, the federal government formed the Environmental Protection Agency (EPA) to create and enforce emissions standards in all 50 states. In response to those new federal and state guidelines, automobile manufacturers started to equip their vehicles with simple emissions equipment beginning with the 1970 model year.

The first piece of emissions equipment was a simple charcoal canister that collected hydrocarbon vapors. The exhaust gas recirculation (EGR) system that routed spent exhaust gases back into the combustion chamber was also introduced early in the 1970s, and it reduced tailpipe emissions of nitrogen oxides. Those mechanical and vacuum-operated systems provided no feedback to tell the vehicle or the owner if they were working. In 1975, the first catalytic converters that significantly reduced the emissions of hydrocarbons and carbon monoxide were introduced. At the time, emissions tests only analyzed tailpipe emissions. Samples of the vehicle's exhaust were tested for particular pollutants, including unburned hydrocarbons, nitrogen oxides, carbon monoxides, and carbon dioxides.

Automotive On-Board Diagnostics

The fuel crisis of the 1970s woke up the automotive world, and the automobile industry realized that fuel efficiency was about to become a consumer and a government requirement. In response to the fuel crisis, the United States Congress established Corporate Average Fuel Economy (CAFE) standards, with which all automotive manufacturers are required to comply. The classic design of the internal combustion automobile engine had gone essentially unchanged from the early 1900s, and it had reached a level of diminishing returns for fuel economy advancements. And it became abundantly clear that the days of an open-loop-style engine were over. At the same time, computer controls, as well as sensor technology, were coming of age. By marrying electronics, sensors, and software to create automotive computer-controlled systems, the fuel efficiency race was on.

With the advent of those computer-controlled systems in the automotive world, the idea of being able to diagnose the performance characteristics of individual components, including emissions equipment, was also born. Prior to 1980,

California Air Research Board

The CARB was established in 1967 in the Mulford-Carrell Act, which combined the Bureau of Air Sanitation and the Motor Vehicle Pollution Control Board. The Air Resources Board (ARB) is a department in the California Environmental Protection Agency. Because its ARB preceded the federal Clean Air Act, California is the only state permitted to have such a regulatory agency. Other states are permitted to follow the ARB standards, which often impose stricter limitations and higher standards, as well as comply with federal regulations.

Currently, CARB works very closely with the EPA, and the EPA relies heavily on CARB's input when developing emissions standards for on-road and off-road vehicles. CARB also certifies non-OEM replacement equipment to ensure that it meets OBD-II system standards and government emissions standards through its Executive Order (EO) program. Owners and technicians should always look for the CARB EO certification label on any equipment being installed on a vehicle. This ensures that the vehicle remains within emissions standards. The major goals of CARB are:

- Provide safe, clean air to all Californians
- Protect the public from exposure to toxic air contaminants
- Reduce California's emission of greenhouse gases
- Provide leadership in implementing and enforcing air pollution control rules and regulations
- Provide innovative approaches for complying with air pollution rules and regulations
- Base decisions on best possible scientific and economic information
- Provide quality customer service to all ARB clients

CARB's website can be visited at www.arb.ca.gov.

CHAPTER 1

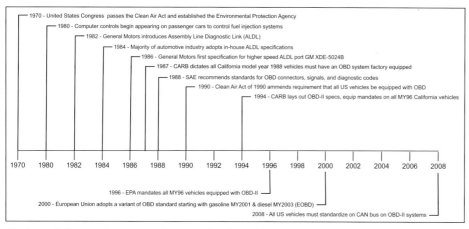

On-board diagnostic systems have evolved over time as electronics have become faster, more capable, and inexpensive. This timeline shows the development and milestones in OBD development.

ECMs were dedicated to individual vehicles, and mainly used to control rudimentary electronic fuel-injection systems in an open-loop-style system. But that was about to change.

Proprietary OBD: 1980–1987

Starting in 1980 on California-bound vehicles, and on all vehicles in the 1981 model year, General Motors equipped its vehicles with its own proprietary ECMs, which controlled the engine as well as emissions equipment. Ford, Chrysler, Honda, and Nissan introduced their own proprietary diagnostics in 1983. A diagnostic port inside the passenger compartment was included. This diagnostic port was originally called the assembly line communications link (ALCL) and was later renamed the assembly line diagnostics link (ALDL).

At the time, scan tools capable of reading data from this diagnostic port were only available to assembly-line workers to ensure that the ECM was functioning properly. When the ECM detected a fault, a Check Engine light (CEL) illuminated on the dashboard. To determine what fault the ECM had detected, the ECM was put into a polling mode—with the ignition on and the engine off, pin A and pin B were shorted together with a wire or paper clip (an early diagnostic tool). The specific two-digit codes blinked in sequence on the CEL light. The technician looked up this code in a dealership's technical manual to determine the fault illuminating the CEL light.

As the complexity of the automobile has increased, placing all of the automotive controls in a single ECM has become impossible. There are too many tasks and too much incoming data from a variety of sensors for one processor to handle. As a result, the automobile now includes a network of multiple control modules that communicate with one another. Newer vehicles have an entire network, much like that of a modern office building with its networked computer systems. As newer technology for the automobile is released, the complexity, the number, and the speeds of the control modules in the automobile continue to increase.

From a scan data standpoint, this ALDL interface was extremely slow when transferring data to the technician. Transmission rates were at a glacial-like 160 baud (bits per second), and that meant a scan tool received data at very a slow rate. Therefore, diagnosing a problem in real time was next to impossible unless it was a static problem. Here's an example of how slow this data rate was: A car going down the quarter-mile while scanning data, including throttle position, RPM, injector pulse, and timing, gave 10 complete sets of data points in a 14-second run. This little amount of data

Although a professional tool was available to trigger diagnostic codes, the common paper clip was commonly used and was known as the generic diagnostic code reader. A paper clip was used to short pins A and B, and this placed the ECM into polling mode. The code reader then interpreted the codes. (Photo courtesy Jim Daley Westech Automotive)

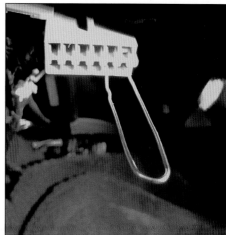

A paper clip is used to short together the two pins on the OBD-I connector. By shorting the pins together, the trouble codes can be read via the blinking dash light. (Photo courtesy Jim Daley Westech Automotive)

AUTOMOTIVE DIAGNOSTIC SYSTEMS

INTRODUCTION TO ON-BOARD DIAGNOSTICS

was useless for determining what was occurring in real time. In late 1985, the standard transfer speed for communication was updated to 8,192 baud, making it possible to collect good scan data information while the vehicle was idling or driving. This made diagnosing non-static faults much easier.

Also at this time, each vehicle manufacturer defined its own connector style, diagnostic codes, and location of the diagnostic connector. The sidebar "General Motors versus Honda OBD Codes" on page 10 shows an example of General Motors' and Honda's diagnostic codes. Notice that not only do they differ, but there was no consistency in the rules, which led to a lot of confusion for independent shops and owners. For example, General Motors #14 is for a coolant sensor fault and Honda #14 is for an IAC fault. And General Motors assigns the same code number for different faults on different vehicles!

In 1987, CARB issued a ruling that all vehicles sold in California would be required to be equipped with a rudimentary OBD system to diagnose engine malfunctions and emissions equipment malfunctions. At this time, CARB did not standardize connector style, connector location, or fault codes. But, this was the foundation for building the OBD (also known as OBD-I) standard.

OBD-I

Thanks to the rulings by CARB in 1968 that an on-board diagnostic system must be included on all vehicles weighing less than 14,000 pounds sold in California, for model year 1987 (MY1987), the Society of Automotive Engineers (SAE) in 1988, and the International Organization for Standards (ISO) in 1989, delivered standards that were known as OBD. Later, after OBD-II was introduced, the previous standards were referred to as OBD-I.

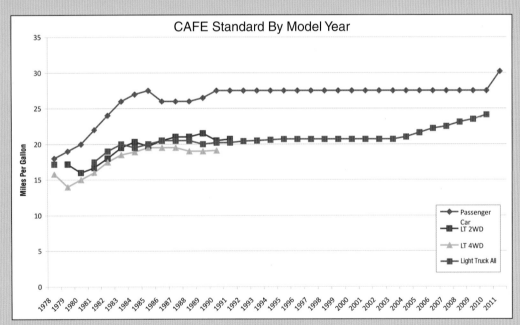

Corporate Average Fuel Economy Standards

As vehicles continue to improve with their computer controls, the CAFE rating continues to increase. (Courtesy Environmental Protection Agency)

Starting in 1975, in response to the Arab oil embargo of 1973, the United States Congress enacted the CAFE standards in order to conserve energy by reducing fuel consumption in passenger cars and light trucks. The National Highway Traffic Safety Administration (NHTSA) is responsible for establishing those fuel efficiency standards, and the EPA is responsible for measuring the fuel efficiency of the vehicles. The first model year affected by the CAFE standards was MY1978.

The standards are established by model year with a targeted CAFE value set for passenger cars or light trucks. Initially, light trucks were distinguished as two-wheel drive and four-wheel drive separately, but those have since been combined.

The CAFE calculation is not simply an average of all of the vehicles that a manufacturer makes. Instead, it also considers the number of vehicles sold.

General Motors and Honda OBD-I Codes

When OBD-I appeared on vehicles, each auto manufacturer used different DTCs (diagnostic trouble codes), and therefore different errors were assigned to the same code. This created massive confusion throughout the automotive service industry, and eventually all manufacturers adopted a standardized set of codes. As a result, OBD-II with its standardized DTC table replaced OBD-I, and all the manufacturers had to comply with this new system. Here are some examples of the General Motors and Honda OBD-I diagnostic codes.

Code	General Motors	Honda
0		ECM normal
1		O_2 Sensor A
2		O_2 Sensor B
3		MAP Sensor
4		Crank Position Sensor
5		MAP Sensor
6		ECT Sensor
7		Throttle Position Sensor
8		Top Dead Center Sensor
9		#1 Cylinder Position Sensor
10		IAT Sensor
11		
12	ECM normal	Exhaust Recirc. System
13	O_2 Sensor open	BAP Sensor
14	Coolant Sensor high	IAC valve or bad ECM
15	Coolant Sensor low	Ignition Output Signal
16	DIS fault	Fuel Injector
17	Cam Sensor fault	Vehicle Speed Sensor
18	Crank/Cam Sensor fault	
19	Crank Sensor fault	A/T Lockup Solenoid
20		Electric Load Detector
21	TPS out of range	VTEC Control solenoid
22	TPS Sensor low	VTEC Pressure solenoid
23	IAT Sensor low	Knock Sensor
24	VSS Sensor fault	
25	Air Temp Sensor fault	
26	Quad Driver #1 fault	
27	Quad Driver 2nd gear fault	
28	Quad Driver #2 fault	
29	Quad Driver 4th gear fault	
30		A/T FI signal A or B
31	Cam Sensor fault	
32	EGR Circuit fault	
33	MAP Sensor high	
34	MAP Sensor low	
35	IAC Circuit fault	
36	Ignition System error	
38	Brake Input Circuit fault	
39	Clutch Input Circuit fault	
41	Cam Sensor fault	Heated O_2 Sensor
42	EST Circuit grounded	
43	Knock Sensor	Fuel System Supply
44	O_2 Sensor lean	
45	O_2 Sensor rich	Fuel Supply Metering
46	Pass-key II fault	
47	PCM-BCM Data Circuit	
48	Misfire diagnosis	Heated O_2 Sensor
51	Calibration error	
52	Engine Oil Temp Circuit low	
53	Battery voltage error	
54	EGR failure/Fuel failure	
55	A/D converter failure	
56	Quad Driver module #2	
57	Boost Control problem	
58	Anti-theft fuel enable	
59		
60		
61	A/C performance/Bad O_2	Front Heated O_2 Sensor
62	High Engine Oil Temp	
63	Right O_2 open	Rear Heated O_2 Sensor
64	O_2 right side lean	
65	O_2 right side rich	Rear O_2 Sensor Heater
66	A/C Pressure low	
67	A/C Clutch failure	Catalytic Converter System
68	A/C Compressor Relay	
69	A/C Clutch Circuit high	
70	A/C Refrigerant high	Automatic Transaxle
71	A/C Evap Temp low	Misfire Cylinder #1
72	Gear Selector Switch	Misfire Cylinder #2
73	A/C Evap Temp high	Misfire Cylinder #3
74		Misfire Cylinder #4
75	Digital EGR #1 Solenoid	Misfire Cylinder #5
76	Digital EGR #2 Solenoid	Misfire Cylinder #6
77	Digital EGR #3 Solenoid	
78		
79	VSS Signal high	
80	VSS Signal low	Exhaust Recirc. System
81	Brake Input Circuit fault	
82	Ignition Control 3x error	
83		
84		
85	PROM error	
86	Analog/Digital ECM error	Coolant temperature
87	EEPROM error	
89		
90		
92		Evap Control System
93		
94		
95		
96		
97		
98		
99	Power Management	

INTRODUCTION TO ON-BOARD DIAGNOSTICS

Some system requirements included:

- Malfunction Indicator Light (MIL) on the dashboard to indicate a malfunction, which can affect emissions
- Any detected malfunction codes must be stored in memory
- Monitoring the catalytic converter system and alert below 60-percent efficiency
- Engine misfire monitoring, storing code and blinking the MIL at a rate of one flash per second
- Monitoring of the evaporative emissions system
- Monitoring of the secondary air injection (AIR) system
- Monitoring of the air conditioning system in regard to refrigerant loss (pre-CFC12)
- Monitoring excess fuel trimming causing excessive rich or lean conditions
- Monitoring oxygen sensor performance and heater circuits
- Monitoring EGR system

The requirements were primarily based on emissions equipment and emissions rules. Regrettably, many of the parameters that can cause emissions problems and trigger the MIL weren't required to be tracked. As a result, it became somewhat difficult to completely diagnose an issue with only partial information available.

At that time common standards did not exist. Therefore connectors, pin locations, error codes, and bus communications varied considerably among manufacturers. A repair facility or home enthusiast needed a scan tool that could adapt to myriad connector types, as well as different data sets. Some systems still used the "blinky codes," while other systems required a scan tool to read any trouble codes that the system sets. Furthermore, each manufacturer could decide what systems it wanted to track for error conditions and what the codes were that were set when an error occurred. Even the

Because no standards existed for the OBD-I connector, each manufacturer was free to use any style of connector. The connectors shown here represent a portion of the OBD-I connectors that were used. (Photo courtesy Jim Daley)

SAE International

SAE International was originally founded in 1905 as the Society of Automobile Engineers by 30 charter-member engineers and a handful of officers that included Henry Ford as its first vice president. It was a result of the world's automobile manufacturers' and engineers' need to gain and share technical information, and to not waste money re-inventing the wheel. In 1916 other societies, such as those for marine and emerging aeronautical technologies, were welcomed to form a professional organization for automotive (self-moving-technology) engineers.

Over the decades, its earlier focus on establishing international engineering standards grew to include establishing itself as an esteemed technical resource and information forum. SAE International and its multiple offshoots and affiliates now involve about 120,000 society members. Though typically engineers and technical specialists, from engineering students to professionals, members are also students and professionals in related academic and public-sector service areas. Membership gives access to not only existing standards, but also upcoming standards and new developments in the industry. Details on the SAE may be found at www.sae.org.

International Organization for Standardization

The ISO is the world's largest developer and publisher of international standards. ISO is a network of the national standards institutes of 163 countries, one member per country, with a central secretariat in Geneva, Switzerland, that coordinates the system. ISO is a non-governmental organization that forms a bridge between the public and private sectors. On the one hand, many of its member institutes are part of the governmental structure of their countries or are mandated by their government. On the other hand, other members have their roots uniquely in the private sector, having been set up by national partnerships of industry associations.

diagnostic lights on the dash indicating a fault had occurred looked different and were called different names. All of this led to a lot of confusion when trying to diagnose problems on different vehicles.

Overall, the initial implementation of OBD was considered a failure from a diagnosis standpoint, although some data was better than no data at all. With too many inconsistencies among vehicles and manufacturers, and the lack of readily available documentation, a new standard was needed.

OBD-I.5

In 1994, General Motors adapted what is termed in the field OBD-I.5 (General Motors never recognized it as such), a small step toward the upcoming OBD-II standards. The 8,192-baud communication was still slow, but it did use the new OBD-II–style connector SAE J1962, which was to be implemented in the upcoming MY1997. The "blinky codes" function was removed, and the pinout for the connector was changed. Dealership or repair shop technicians could flash the memory and reprogram the ECMs. This meant that the software in the ECM could be written through the ALDL connector instead of having to change an electrically erasable programmable read-only memory (EEPROM) chip inside the ECM.

Unfortunately, the trouble codes were still based on a two-digit system akin to that used in the OBD-I scheme. At this point, there still was no standardization for trouble codes from an accredited organization, such as CARB, ISO, or SAE. This still meant that a multitude of reference materials had to be kept on-hand to help diagnose the meaning of the numeric trouble code for each platform.

Another first in the OBD-1.5 system incorporated other vehicle systems in the diagnostic tests. Those systems included entertainment systems; heating, venting, and air-conditioning (HVAC) systems; active handling systems; passive restraint systems; and safety systems such as tire pressure monitors. At that time, electronically controlled transmissions were also being installed on new vehicles. The ECM could be responsible for controlling this too, as well as performing the diagnostics for it.

Take care when identifying this hybrid OBD-1.5. Certain models used the OBD-I–style connectors, while later systems used the OBD-II SAE J1962-style connector. A quick check of the emissions sticker under the vehicle's hood confirms the level of OBD that the vehicle supports. If an OBD-II scan tool is hooked up to an OBD-1.5–equipped vehicle, it may or may not successfully communicate with the vehicle. The power connections in the connector are shared by OBD-1.5 and OBD-II, so neither system should harm the scan tool if it is accidentally connected to the incorrect system version.

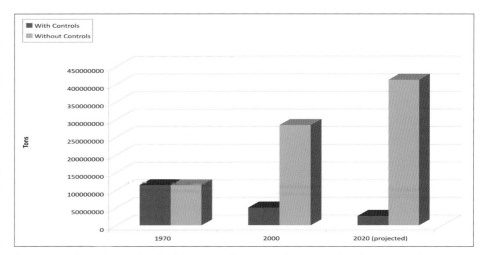

Shown are the effects of emissions control systems in terms of carbon monoxide emissions. (Courtesy Environmental Protection Agency).

OBD-II

In 1994, CARB and the EPA realized that there were many downfalls with the current manufacturers' OBD systems. The EPA amended the Clean Air Act of 1990 to include the requirement that all vehicles sold in the United States be equipped with some type of OBD system. CARB worked with the EPA to establish the rules for an OBD system that included standardized fault codes, connector locations, connector pinouts, information on the data bus, etc.

The OBD-II specifications are constantly evolving as fuel economy requirements increase, emissions standards continue to change, and electronic hardware continues to perform better at lower cost. Thanks to heavy involvement of ISO and SAE, the rules for OBD-II have been well defined. Each original equipment manufacturer (OEM) must now comply with OBD-II standards instituted by ISO and SAE. The benefits of developing OBD-II are not limited to ensuring that vehicle systems are performing within specifications. Standardization also made it easy for a technician or home enthusiast to read and understand a single set of information when diagnosing a potential system failure.

Has OBD Made a Difference?

According to the EPA, the emissions control capabilities of OBD systems have dramatically decreased airborne pollutants. The number of vehicles currently on the road in the United States has increased every year, except for a slight reduction in 2009. Even though there are more vehicles putting more pollutants into the air, research shows that there is still a dramatic decrease in emissions.

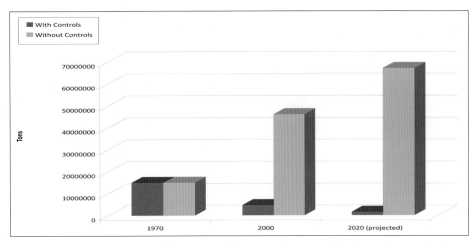

Shown are the effects of emissions control systems in terms of hydrocarbon emissions. (Courtesy Environmental Protection Agency)

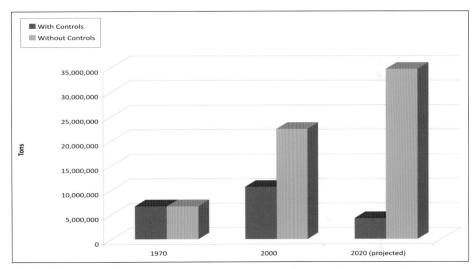

Shown are the effects of emissions control systems in terms of nitrogen oxide emissions. (Courtesy Environmental Protection Agency)

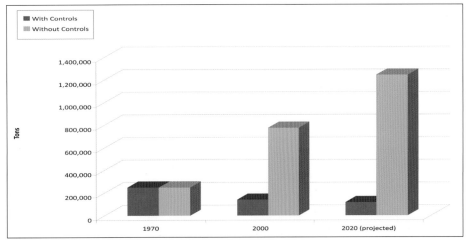

Shown are the effects of emissions control systems in terms of fine particulate emissions. (Courtesy Environmental Protection Agency)

CHAPTER 2

OBD-II STANDARDIZATION

The standards that became OBD-II were cooperatively developed by CARB, SAE, ISO, and EPA. With those standards, automobile manufacturers had new guidelines to follow when developing their implementation of OBD-II in the engine management control systems. No longer were mechanics faced with cryptic codes, blinking lights, paper clips, strange connections, and no instructions on what any of the diagnostic feedback meant. As a result, repair shops and home enthusiasts have since benefited from a market well supplied with scan tools able to communicate with multiple manufacturers' ECMs.

In a nutshell, OBD-II's main goals are:

- Increase fuel economy by ensuring optimal engine running conditions
- Reduce emissions
- Reduce the time between a system failure and notifications through constant monitoring and comparison of acceptable system data
- Aid in the diagnostics and repair of emissions equipment

OBD-II also went beyond the basic diagnostic capabilities of OBD-I by including operations outside of mere system pass/fail protocols, including the ability to do the following:

- Capture and display current engine/system conditions in real time
- Store diagnostic trouble codes (DTC) in non-volatile memory to be recalled later
- Store any pending DTCs that have not yet tripped the MIL
- Show data captured at the instance a DTC code was established (freeze-frame data)
- Ability to clear any DTCs that have been set via a scan tool
- Store and display system readiness monitors (SRM) via the system readiness tests (SRT)
- Poll information about the vehicle (from the ECM) regarding vehicle identification number (VIN), model, engine, transmission, etc.
- Allow real-time controls to test a variety of engine management and transmission management systems via a scan tool.

The Power of the Microchip

More than 30 years ago, the co-founder of Intel (the microchip manufacturer), Gordon Moore, boldly predicted that microchip development would double the number of transistors on a microchip every 24 months. He was right; the number of transistors has exponentially grown over time, allowing the industry access to faster processors and more data. As a result, the capabilities of the automotive-industry ECM have increased, although not at the accelerated rate of the microprocessor market.

The OBD-II–equipped vehicles represented a major upgrade in computer hardware capabilities and storage capacity. The software driving the ECM was able to be more robust, which meant that it could be expanded with additional sensor capabilities, more tables for data lookup, and more diagnostics. As processors are becoming faster, the traditional lookup tables for data points are being replaced by real-time, equation-driven intelligence, which adapts to a wider variety of situations. The industry has seen the microprocessor controllers for the automobile transition from a simple 8-bit processor to a complex 64-bit multi-tasking processor that communicates to numerous other processor based modules located throughout the entire vehicle. The communication between systems within the vehicle has become so complex that vehicles now have an entire network, with high-level network structures and languages.

OBD-II STANDARDIZATION

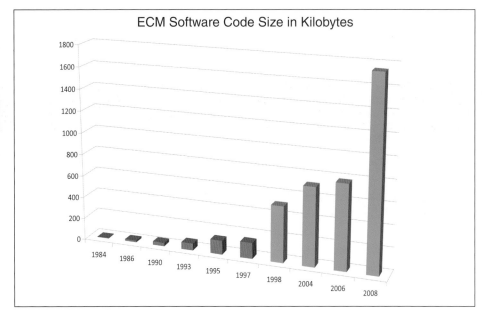

As ECMs gained power and capacity, the capabilities of the software driving them also increased. OBD-I ECMs are blue. OBD-II ECMs are green.

As automobile complexity increased, placing all of the automotive controls in a single ECM became impossible. So the automobile has become a network of multiple control modules that communicate with one another. Newer vehicles have a vehicle-wide network, much like that of a modern office building with its computer systems. And as newer technology for the automobile is released, the complexity and speeds of the control modules within the automobile continue to accelerate.

Evolution of Automotive Networks

The initial communication within the automobile system consisted of a single control point in the ECM. This controlled the various automotive performance subsystems, such as the transmission and emissions equipment, while communicating with other stand-alone systems, such as the anti-lock braking system and climate control system. The system's architecture is built in a ring fashion. Consequently, each component or sensor puts data on the bus, and this data is transferred serial from component to component until it winds its way back to the ECM that requested the information. The system's architecture processed slowly because the information had to travel through the entire network and then return to the ECM. Thus, speeds inside this network topology could be so slow that real-time data acquisition did not yield enough meaningful data to diagnose an issue. Furthermore, a network topology of this scheme required a lot of wiring because the network was literally point-to-point wiring. As vehicle weight became an issue in terms of fuel economy, reducing hard wiring and the mass of the wires became an important issue for engineers. It became quickly apparent to automotive engineers that a ring topology would no longer suffice for the automotive network.

A new topology, sometimes referred to as a star network, was implemented in most new vehicles. The structure is such

A mid-1980s General Motors ECM with the Memcal installed. This ECM was only capable of controlling rudimentary engine controls, such as ignition timing and fuel injection timing. (Photo courtesy Jim Daley Westech Automotive)

This is an early 2000s General Motors ECM from a Camaro. This ECM not only controls all engine and emission features, but it also controls the electronic transmission and communicates with the body control module and anti-lock brake module.

CHAPTER 2

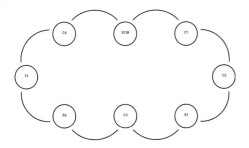

A ring topology for a network consists of a point-to-point connection between sensors and controllers. All the controllers pass the information around the ring. The issues in this system are latency, the amount of wiring required, and any break in physical connections, which can shut down the entire system.

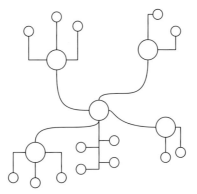

A star topology breaks down the components and systems into small clusters and as a result, only passes data to the master controller. With star topology network, latency is reduced and the system's overall reliability is significantly increased.

that each system is considered a node on the network. Sub-networks responsible to their master node can be queried and controlled without having the rest of the network see and pass through commands that have nothing to do with the other systems. As a result, speeds increased significantly, as well as the amount of data that could be passed between the systems. Furthermore, the complexity of the systems increased as a single ECM was no longer responsible for complex controller systems, such as ABS, multimedia, digital dashboards, etc. And automotive manufacturers incorporated more-advanced vehicle technology, such as active suspensions, passive safety systems, increased emissions monitoring, and redundancy in critical vehicle systems.

As hardware and networking technology continued to evolve, automotive networks benefited from those changes in technology. More vehicle systems and sensor complexity require faster data transfer rates, as well as a significant increase in data traffic. The ECM no longer manages all of the vehicle's performance and emissions aspects. And with advances in hardware, vehicle sub-systems now have their own controllers, which synchronize with all the other controllers in the vehicle. Many newer vehicles can have 10 or more controllers, all communicating with one another and responding to any external requests from scan tool monitors.

Standardized ALDL Connector

As discussed in Chapter 1, most OEMs chose their own diagnostic connector style, even varying it among their different models. This meant that any scan tool used to diagnose OBD faults required multiple connectors to accommodate a variety of different connector styles on different vehicles. On vehicles that did not support code reading and instead relied on blinking codes, the pins required to activate the blinking code varied. Shorting the wrong pins together

The SAE J1962 specification defined a standard shape and pin location for the OBD-II connector.

to read codes could not only potentially harm the ECM, it could also cause injury to the technician and/or cause an electrical fire.

The SAE released SAE-J1962 standards, which relate specifically to the style and shape of the OBD-II connector. This connector, known as a Type-A connector, is meant for vehicles with a 12-volt system. For vehicles with a 24-volt system, a similar connector is used with a different plastic center divider. This divider ensures that the mechanic can't damage the scan tool by connecting a 12-volt scan tool to a 24-volt system.

Standardized Scan Tool Data

In the formative years, manufacturer-specific protocols for on-board diagnostic systems varied significantly, between manufacturers and even between vehicles. Communication speeds varied from 160 baud to 8,192 baud. It is very difficult to diagnose a vehicle using real-time data at very slow scan speed; the data cannot keep up with the conditions. For example, when trying to diagnose an early OBD-I vehicle, the scan tool may only receive one or two complete data points per second, which can make it difficult to catch a quick glitch of a sensor or get a smooth average of readings from another sensor. Also, scan tools for OBD-I had to cover multiple language protocols. This made it difficult and expensive for professional shops to have access to scan tools that covered all the different vehicles that came in for diagnosis, and it confused home mechanics about the compatibility of scan tools.

OBD-II was designed to standardize the communication of scan tool data. The SAE J1978 specification standardized the methodology of communication between an OBD-II system and an external scan tool. The transfer rates of scan tool data are set at 500,000 baud

OBD-II STANDARDIZATION

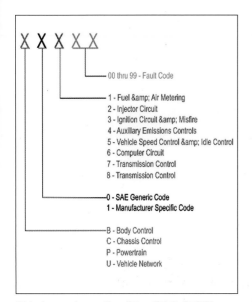

This is a schematic of the OBD-II DTCs. Each field differentiates the type and system of the DTC.

(or 500 kilobytes per second). Considering this was a magnitude of times faster than the OBD-I systems, real-time scan data could now be used to help diagnose issues. A mechanic could view real-time data points 10 times per second or more, so he or she could see and understand what was occurring and help diagnose the problem.

Another item addressed for standardization was the format of the diagnostic data. This allowed one common scan tool format for the variety of manufacturers. A professional shop no longer needed several different scan tools to scan incoming vehicles. Now, a single scan tool covered all of the incoming OBD-II–equipped vehicles using a common, simple software interface. This also opened up the market for inexpensive hand-held scanners, which were available to the home mechanic with the same easy-to-use interfaces.

Diagnostic Trouble Codes

As discussed in Chapter 1, prior to OBD-II, each manufacturer was free to determine which systems were diagnosed and its own DTCs. To illustrate this point, refer to "General Motors versus Honda OBD Codes" on page 10 in which General Motors listed code #75 as a bad EGR valve and where Honda used the number-75 code to mean there is a misfire indicated on cylinder number-5 ("cylinder #5" in OBD code). This resulted in a lot of confusion and difficulty in diagnosing specific information reported from the OBD system because there were no standards on fault reporting.

This all changed with the advent of OBD-II. The DTCs were standardized according to SAE J2012. This specification defined the DTC code as having five digits. The DTC digits specifically give information about where the fault is generated, what type of fault it is, and the specific fault itself. Today, simply decoding the 5-digit code informs the technician exactly what the DTC code indicats regardless of the make or model of the vehicle. For example, a standard DTC for a general engine misfire is P0301; "P" for is Powertrain, "0" indicates that it is a standard SAE code, "3" denotes it is related to ignition or misfire, and "01" indicates a misfire on cylinder number-1.

Most new scan tools have an excellent library of generic diagnostic codes that are specified by the OBD-II standards as well as manufacturer-specific diagnostic trouble codes. This eliminates any need to memorize all of the trouble codes or to have a service manual available to decode them. The internet is also a good source for decoding trouble codes if the scan tool does not have an on-board DTC library.

The MIL

The MIL (aka Service Engine Soon light) has been around since the first OBD systems. Initially, in OBD-I systems, the MIL was triggered any time that a DTC was detected. Prior to OBD-II, MILs were located anywhere within the passenger compartment and could simply be an illuminated LED or incandescent lamp. Now, a graphic or word is located directly on the dashboard, and is illuminated in red or yellow. Unfortunately, as of this writing, the shape and wording of the light isn't standardized. Some manufacturers use a picture of an engine, some use the words "check engine," or a variation, and

A DTC is shown for an engine misfire. Many scan tools provide the DTC number and a brief description of the fault code.

A graphical representation of the MIL is shown as a red engine icon on the dash screen.

"Service Engine Soon" in yellow represents the MIL light.

AUTOMOTIVE DIAGNOSTIC SYSTEMS 17

CHAPTER 2

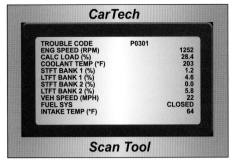

For set DTCs, the scan tool can display run-time data at the time the fault occurred. The amount of data available varies based on the capabilities of the scan tool interface and ECM memory.

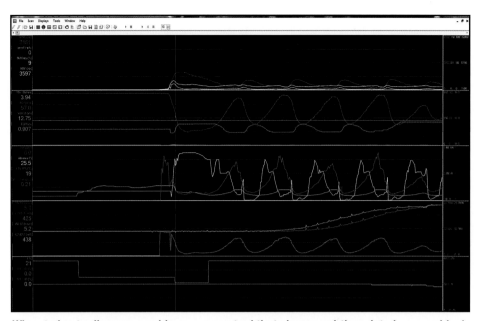

When trying to diagnose problems, a scan tool that shows real-time data in a graphical format is extremely useful. Much of the data from the various sensors in the feedback and control systems changes almost instantaneously. As a result, tracking real-time data without a graphical interface is difficult because these rapid changes may occur quicker than the textual scanner can update. The graphical representation of data allows you to go back and review all the data over time. The graphical representation also allows you to simultaneously check multiple data sets as they occur.

one manufacturer just shows the picture of a wrench.

The variety of lights, icons, and directives can confuse the everyday driver about what the ECM is trying to indicate. When the MIL is illuminated, a common misconception is that the vehicle is no longer drivable or failure is imminent. This is not necessarily the case. Many of the problems that the ECM detects, and that cause it to illuminate the MIL, are due to emissions equipment malfunction or a failed sensor. Luckily, the modern automobile is equipped with failure modes that typically do not leave the automobile stranded or disabled. Typically, when a problem is detected and the MIL illuminates, the vehicle produces poor fuel mileage, lack of performance, and generally exhausts more hydrocarbons than it should. In severe cases, vehicles enter a limp-home mode, which consists of very limited performance but allows the vehicle to get to a dealership or repair facility safely and under its own power.

Frame-to-Frame Data

In order to determine the exact state of the automobile when the fault occurred, the ECM stores a snapshot of the conditions in memory. This data is known as freeze-frame data, and the data correlates to the specific fault that was generated. Depending on the available memory within the ECM, the freeze-frame data can include snapshots before, during, and after the fault. At a minimum, a snapshot of the operating conditions' status exactly when the fault occurred is stored in the ECM's memory. Freeze-frame data can clear itself in certain situations if the trouble code expires and is cleared after meeting certain criteria. Typically, this clearing requirement only occurs after many cold starts, and if the problem has been cleared. This gives ample time for the DTCs to be downloaded and the vehicle to be diagnosed.

Real-Time Data

Another benefit of the OBD-II standardization is the ability to poll the system to read real-time data from sensors and calculations made by the ECM. With the established standards under OBD-II, a common set of values is shared among all manufacturers. These data sets are known as the generic parameter identifications (PIDs), which allow the mechanic to view the information coming from the sensors and the ECM in real time. The mechanic looks for possible invalid data or clues about the current running conditions of the automobile. Furthermore, OEMs also can create their own specific sensors and calculations, which support their applications. However, all scan tools may not be able to support or read these manufacturer-specific codes.

The way a scan tool handles the real-time data determines its functionality and value. A basic, inexpensive, and hand-held scan tool may display real-time data in textual format only. Some scan tools may allow that data to be downloaded to a personal computer

OBD-II STANDARDIZATION

and imported into a spreadsheet program to create a graph. Other scan tools may have their own graphical review software that installs on a personal computer. Higher-end scanners allow text and graphical representation of the real-time data directly on the screen of the scan tool, which makes it much easier to review and diagnose issues from the data right at the vehicle.

What Information Does OBD-II Provide?

OBD-II is much more than just the MIL and the DTC. Underneath the specifications is a plethora of information regarding the communication between vehicle modules, emissions controls, etc. For the mechanic, it's essential to understand and use the diagnostic information to determine the status and condition of the vehicle.

A common misconception is that OBD-II DTC codes indicate exactly what is wrong with the vehicle. Imagine going to the doctor with a sore throat. Typically, a doctor does not offer a diagnosis and write a prescription based on the single symptom. Instead, he or she will take that symptom into account, perform more tests, and come up with a confident diagnosis and cure for the sickness. A mechanic will take the same path. The DTCs inform the mechanic that certain systems are not performing within their predetermined specifications. Many times, as highlighted throughout this book, the ultimate problem or root of the problem may be causing other systems to fall out of spec. The other symptoms lead the mechanic to look at incorrect solutions. Just like the example of the doctor and the sore throat where the sickness may actually be in the patient's ears, the mechanic should use the data provided by the entire OBD-II system to examine the entire closed-loop system.

SAE Standards versus ISO Standards

SAE developed and published many of the original OBD-I and OBD-II standards. Because the automobile industry is a global and international industry, and a majority of the world follows the ISO, the original SAE specifications have been rolled into ISO specifications. Figure 2.1 shows the conversion from the SAE standard to the ISO 15031 standard. For example, SAE J1930 has been incorporated into ISO 15031.2, while SAE J2012 has been incorporated into ISO 15031.6. Referring to either standard is acceptable because they cross reference to each other. A quick browse of the SAE website reveals a lot of information on the ISO specifications, as well as cross references.

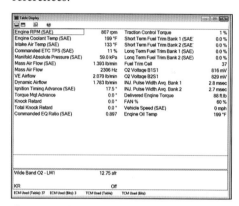

A scan tool can show real-time data in a text format. Although useful for looking at instantaneous data, it is difficult to analyze an overall picture of the data. Most text screens on scanners have a refresh rate that runs much slower than the data is changing. So some data may not be shown as it has occurred because it has changed before the scan tool screen could be updated. Also the human eye can only detect and process data at a certain rate. Unfortunately, the change in the data can far exceed the rate that the human brain can process the data.

SAE Standards

J1930	J1962	J1978	J1979	J2012	J2186
Determines abbreviations, definitions and acronyms	Defines physical and electrical standards for interface connector	Defines the standards for scan tool interfaces	Defines the diagnostic test modes	Defines the diagnostic trouble codes (DTC)	Defines the data link security and seed keys
2	3	4	5	6	7

ISO 15031

Figure 2.1. Multiple SAE standards are incorporated into a single ISO standard. Many times, the standards are quoted interchangeably due to their overlap.

CHAPTER 3

THE OBD-II DATA INTERFACE

The OBD-II data interface is the main conduit used to query diagnostic information from the vehicle's on-board systems, read real-time data on running conditions of the vehicle, download stored diagnostic data, and, in some cases, program the on-board systems. The interface is accessed through a standardized connector located in the passenger compartment. Communications to the interface are specified through SAE and ISO standards. The OBD-II interface powers most scan tools because it draws from the vehicle's OBD-II 12-volt electrical system and a ground present in the connector. The connection to the scan tool typically requires the ignition key to be in the run position, but the vehicle does not necessarily have to be running when the scan tool is active.

Data Link Connector

The scan tool physically interfaces with or connects to the data link connector (DLC) to access the OBD-II system. The DLC is a 16-pin D-style female connector defined by the specification SAE J1962 (and its European counterpart, ISO/DIS 15301-3). This particular specification dictates the physical location of the connector within the vehicle, the shape and size of the connector, and the electrical connections/pinouts in the connector. There are two different versions of the J1962 connector: Type-A, which is for 12-volt-equipped vehicles, and

J1962 Pin Configuration

1. General Motors: J2411 GMLAN/SWC/Single-Wire CAN (or manufacturer's discretion)
2. SAE-J1850 PWM and SAE-J1850 VPW (+) positive signal
3. Ford DCL (+) positive
4. Chassis ground
5. Signal ground
6. SAE-J2284 and ISO 15765-4 CAN signal high
7. ISO 9141-2 and ISO 14230-4 "K" line
8. No Connect
9. No Connect
10. SAE-J1850 PWM (-) negative signal
11. Ford DCL (-) negative
12. No Connect
13. No Connect
14. SAE-J2284 and ISO 15765-4 CAN signal low
15. ISO 9141-2 and ISO 14230-4 "L" line
16. Battery voltage

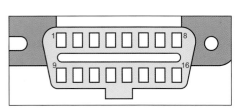

Type-A SAE J1962 connector. Note the single center section that defines the voltage as 12 volts.

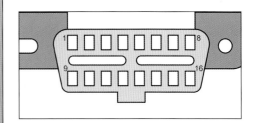

Type-B SAE J1962 connector. Note the split center section that defines the voltage as 24 volts.

Type-B, which is for 24-volt-equipped vehicles. The main difference between the 12-volt version and the 24-volt version is the middle divider on the connector. By splitting the connector on the 24-volt version, the 12-volt scanner cannot be accidentally hooked to a 24-volt system, thus preventing an accident that may damage the scan tool interface. If there is a doubt, use a volt/ohm meter (VOM) to measure the voltage at the DLC to determine the voltage on the OBD-II system.

THE OBD-II DATA INTERFACE

In most situations, the DLC is located under the dashboard, near the steering wheel, and easily visible.

SAE J1962 also specifies the location of the data link connector. Per specification, "the DLC shall be located within a 3-foot sphere from the driver's position, and be located within the passenger compartment." The DLC may be out in the open behind an interior panel, behind the ashtray, or beneath an access hatch. It may or may not have a cap over the plug. Furthermore, by specification, no tools should be required to access the connector. In the event that the connector is in a non-standard location, the OEM includes a sticker on the underside of the hood to denote the non-standard location.

It is important to note that the J1962 connector has a pronounced "D" shape along with a key on the short side of the connector. The connector on the vehicle has this same unique shape, and is designed so that the connector cannot be placed upside down on the vehicle. But I've witnessed a few technicians forcing the connector onto the vehicle upside down. It can be done if enough force is applied because the components are made of plastic and can distort enough to do this, but you need to avoid making this mistake. Therefore, take care to note the orientation of the connector before plugging in the scan tool cable.

Determining if the Vehicle is OBD-II

Unfortunately, the style of the connector alone does not determine if a particular vehicle is OBD-II equipped. Many

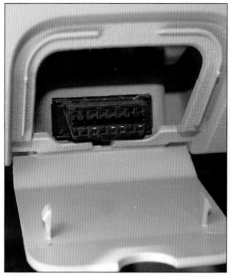

The DLC is located under an access panel in some vehicles. Typically, this access panel is clearly labeled and should be removable without any tools.

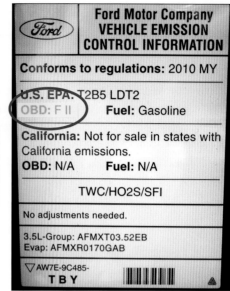

The underhood sticker provides verification of an OBD-II-equipped vehicle. The yellow highlighted area indicates OBD-II.

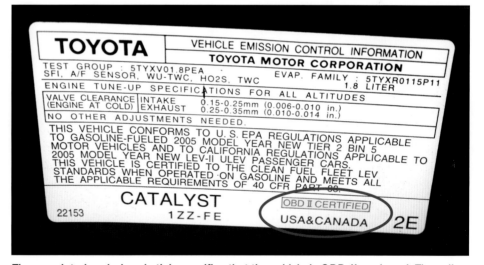

The mandated underhood sticker verifies that the vehicle is OBD-II equipped. The yellow highlighted area indicates OBD-II specification.

OEM manufacturers began to use the J1962 OBD-II connector as their OBD-I–equipped vehicles were being produced. And a few foreign-built vehicles use their own diagnostic connectors rather than the standard J1962 connector. Reference the emissions sticker located under the hood to determine whether or not the vehicle is OBD-II equipped. By law, this sticker must be located in a non-obstructed area, and it is typically located under the hood.

If a vehicle is OBD-II equipped, it will be noted prominently on the sticker.

Some commercially available scan tools read OBD-I– and OBD-II–equipped vehicles. These scan tools have a variety of adapter connectors in the kit to allow you to connect to a variety of vehicles. *Note:* Refer to the owner's manual if there is any doubt about the capabilities of the scan tool to connect to a particular vehicle's system.

CHAPTER 3

OBD-II Data Protocols

Even though the OBD-II standard is fairly well defined, there are a variety of different protocols or formats in which the data stream is presented through the DLC. Initially, each manufacturer followed a different standard protocol. Unfortunately, those protocols are not compatible with each other, and it is up to the scan tool interface to be able to adapt to the differences in pin configurations, data transfer rates, and signal voltage levels. The only commonality between the interfaces is that OBD-II J1962 specification requires that pins 4 and 5 are connected to ground, and pin 16 is connected to +12v or +24v. It is sometimes possible to look at which pins are present to determine which protocol is used, but usually you have to consult the vehicle's diagnostics book or the scanner configuration.

The methodology with which the ECM communicates with the outside world or other modules is known as the communications protocols. OBD-II supports a variety of communication protocols, which have evolved over time. These protocols include: SAE J1850 VPW, SAE J1850 PWM, ISO 9141-2, ISO 14230 KWP2000, and ISO 15765 CAN.

Serial Communications Protocols

Each of the defined OBD-II protocols implements serial communications. Not only is each pin in the connector defined, the standard also defines the format of the data transmitted by any of the devices communicating on that protocol. The basis for serial communication is that one piece of information is transmitted at a time. Figure 3.1 shows a graphical representation of the data passed along the data bus. In simple terms, think of this packet as a simple file folder. It contains information that not only has the specified data requested, but it also has information that shows when the data starts, when it ends, a check to ensure that the data is valid, and a response from the receiver to ensure that the data has been received accurately.

In the serial stream, each chunk of the signal format is coded. The chunks appear in the order in which they are transmitted.

1. Start of frame (SOF) information is sent out letting the receiver know that there is data coming down the bus.
2. Header information (HI) is sent out telling the receiver what kind of data is to be transmitted, the priority of the data, and whether or not a response from the sender is expected.
3. The data is sent to the receiver. This information is pertinent to whatever is being polled.
4. A cyclic redundancy check (CRC) is sent so that the receiver can compare this value to data to ensure that the data received is valid.
5. The end of data (EOD) symbol immediately follows the CRC information. This lets the receiver know that the transmitted data is now complete.

SAE J1850 Transmission Signal Format

Figure 3.1 This is the signal transmission format for the SAE J1850 plug.

6. If the sender has requested confirmation from the receiver, the in-frame response (IFR) information is sent back to the sender to acknowledge that the transmission was received correctly.
7. Along with IFR, the receiver can also send back a CRC to ensure that the data was sent back to the sender in the IFR correctly.
8. An end of file (EOF) marker designates the end of the entire transmission.

All of this information was just to send one byte of information on the data bus! Imagine the amount of data sent on the bus while scanning a vehicle!

The makeup of the signal is a bunch of segmented data streamed together in a continuous chunk of data. The sender puts the information on the data bus one bit/byte/word/packet at a time and waits for acknowledgement from the receiver. It may seem like this is a very inefficient way to transmit data, but the main advantage is that the data can be sent at very high rates. This is critical because there is a lot of information flowing on a standard automobile's network, and this transmission requires only one or two wires, a significant savings on the amount of wiring required for the automobile's network. The rest of

	SAE J1850 PWM Protocol (Ford)	SAE J1850 VPW Protocol (General Motors)
Data Pins	Pin 2: Data bus low, Pin 10: Data bus high	Pin 2: Data bus high
Logic 1 Voltage	+5v	+3.5V (max +7v)
Transmission Speed	41.6 kilobauds	10.4/41.6 kilobauds
Message Length	12 bytes	12 bytes

THE OBD-II DATA INTERFACE

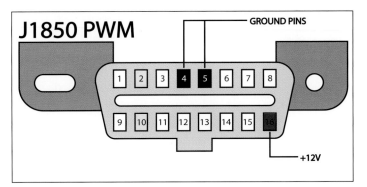

The DLC commonly found on General Motors vehicles uses a variable-pulse-width communications protocol.

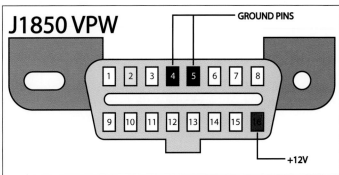

The DLC commonly found on Ford vehicles uses a pulse-width-modulated communications protocol.

the protocols that make up the variety of OBD-II's protocols function similarly to this description.

SAE J1850 Communications Protocol

The SAE standardized the SAE J1850 protocol because its low cost and open architecture enabled it to become the standard for in-vehicle network communications. Two types of this protocol were adopted: GM adopted a variable pulse width (VPW) scheme, while Ford adopted a pulse width modulated (PWM) scheme. The difference between VPW- and PWM-style protocols is beyond the scope of this book but, in lay terms, there are a few simple differences as shown below.

ISO 9141-2 Communications Protocol

Chrysler, Asian, and European manufacturers primarily use the ISO 9141-2 for 2000–2004 models.

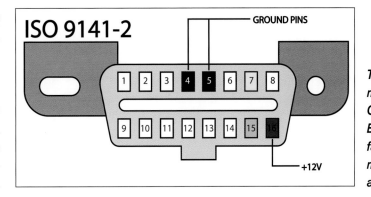

The DLC is commonly found on Chrysler, Asian, and European manufacturers between model years 2000 and 2004.

ISO 9141-2 PWM Protocol

Data Pins	Pin 7: K-line, Pin 15: L-line (optional)
Logic 1 Voltage	+5v (maximum of battery voltage)
Transmission Speed	10.4 kilobaud
Message Length	12 bytes

ISO 14230 KWP2000 Communications Protocol

Asian and European manufacturers primarily use the ISO 14230 KWP2000 (key word protocol 2000) protocol. It is used on Alfa Romeo, Audi, BMW, Citroen, Fiat, Honda, Hyundai, Jaguar, Jeep (post-2004), Kia, Land Rover, Mazda, Mercedes, Mitsubishi, Nissan, Peugeot, Renault, Saab, Skoda, Subaru, Toyota, Vauxhall, Volkswagen (post-2001), and Volvo (pre-2005).

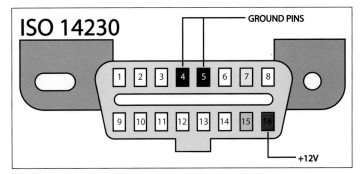

Asian and European manufacturers primarily use ISO 14230 KWP2000 (key word protocol 2000) protocol.

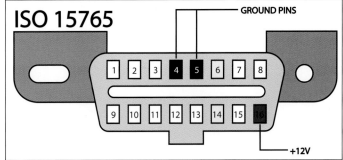

The DLC uses ISO 15765 CAN communications protocol. All vehicles produced after 2008 use this protocol.

AUTOMOTIVE DIAGNOSTIC SYSTEMS

CHAPTER 3

ISO 14230 KWP2000 Protocol

Data Pins	Pin 7: K-line, Pin 15: L-line (optional)
Logic 1 Voltage	+5v (maximum of battery voltage)
Transmission Speed	1.2 to 10.4 kilobauds
Message Length	Up to 255 bytes

ISO 15765 CAN Communications Protocol

The ISO 15765 CAN protocol has been popular outside the United States and U.S. auto manufacturers are slowly implementing it. All vehicles made after 2008 use this communications protocol.

ISO 15765 CAN Protocol

Data Pins	Pin 6: CAN high, Pin 14: CAN low
Logic 1 Voltage	2v difference between high and low
Transmission Speed	250 or 500 kbit/sec
Message Length	Up to 4095 bytes

Protocol	Data Pins	Power/Ground Pins
SAE J1850 PWM	2, 10	16/4,5
SAE J1850 VPW	2	16/4,5
ISO 9141-2	7, 15 (optional)	16/4,5
ISO 14230 KWP2000	7, 15 (optional)	16/4,5
ISO 15765	6,14	16/4,5

Troubleshooting Common DLC Connection Problems

Communication to the OBD-II system is generally considered pretty straightforward. The diagnostic link connection can only be made and implemented in one way, right? But certain

Testing Voltage at the DLC for CAN-Based Vehicles

To determine if the vehicle is using the ISO 15765 CAN communications protocol, the resistance between pins needs to be confirmed. Set the VOM to measure resistance. If it is not an auto ranging style VOM, set the range to allow 60 ohms to be read. Turn the vehicle's ignition to the RUN position. Insert the red/positive lead from the VOM into pin 14 on the DLC. Insert the black/negative lead from the VOM into pin 6 on the DLC.

Observe the reading on the VOM. The reading should be somewhere around the 60 ohms. In this example, the reading is for a 2010 ISO 15765 CAN equipped vehicle. The reading on the VOM indicates that the resistance between the pins is 61.2 ohms, which confirms the CAN protocol on this vehicle.

If no voltage can be measured at either position, then there is a problem with the power that is provided to the DLC connector. Most vehicles share the accessory circuit or cigarette lighter circuit with the DLC. A quick check of the fuses to see if a fuse has blown typically will remedy the lack of voltage.

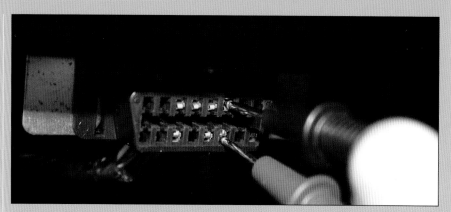

When testing to see if the vehicle is CAN-bus-based or not, use a VOM to test the resistance between pins 14 and 6. The polarity of the probes does not make any difference in the test.

If the vehicle is equipped with a CAN bus structure, the VOM shows a resistance near 60 ohms. If the VOM reads the connection as open ("OL" on most VOMs), then it may be a non-CAN-based vehicle.

THE OBD-II DATA INTERFACE

issues can make connecting and communicating to the system troublesome. These major issues include: ignition-key position, connector-voltage issues, incorrect communications protocols, "hung" ECM (frozen—needs to be re-booted), missing or bad data, and OEM-updated ECM software that is unknown to the scan tool.

Ignition Key not in Run Position

Most OBD-II scan tools require that the ignition key be placed in the run position, or the engine to be running. Also, many vehicles take several seconds for all of the modules to finish their boot-up sequences. A good rule is to engage the scan tool after all chimes have finished and all dashboard activity has been completed. The scan tool may not connect to the OBD-II system if you do not wait for the complete system boot-up.

No Voltage or Low Voltage on Connector

By specification, every OBD-II connector must have 12-v power on pin 16 and ground on pins 4 and 5. The scan tools rely on this power being present. First, the ignition key must be in the run position. Using a voltmeter set to measure DC voltage, place the red/positive lead on pin 16 and the black/negative lead on pin 4 or pin 5. The voltmeter should read 12-v DC (or near to it). If the reading is much below 11 volts, some scan tools have problems connecting. If this is the case, inspect the battery and/or charging system first. If you read no voltage between the pins, a blown fuse is most likely the problem. Typically, the DLC power is shared with the accessory power on the fuse panel. Using the owner's manual for reference, locate the fuse panel and check to see if any fuses have been blown. If so replace the fuse and recheck your power.

Incorrect Communications Protocol

Unfortunately, there are five different protocols that can be used on OBD-II while using the same J1962-style

Testing Voltage at the DLC for Non-CAN-Based Vehicles

To test the 12-volt power to the DLC, first set the VOM to DC voltage. If it is not an auto-ranging-style VOM, set the range to allow 12 volts to be read. Turn the vehicle's ignition to the run position. Insert the red/positive lead from the VOM into pin 16 on the DLC. Insert the black/negative lead from the VOM into pin 4 or pin 5 on the DLC.

Observe the reading on the VOM. The reading should be somewhere around the voltage of the battery system on the vehicle. In this example, the reading is for a 12-volt vehicle. The reading on the VOM indicates that power is present on the DLC and a scan tool can be connected to it.

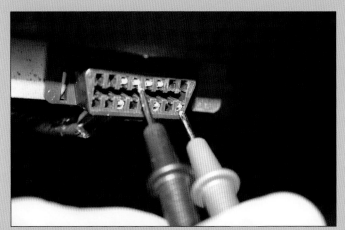

Serial data transfer sends one bit of information after another on a single wire until the transfer is complete. Parallel data transfer sends all the bits of information at once on multiple wires.

If the vehicle is not equipped with a CAN bus structure, the VOM shows the battery or system voltage, which is usually around 12 volts DC. If the VOM reads no voltage, then it possibly may be a CAN-based vehicle.

CHAPTER 3

Ensuring Valid Data Transmissions

When hardware is transmitting data across wires, there always exists the possibility that the data becomes corrupt somewhere between the sender and the receiver. These checks occur in just about any electronic device whether it be automotive systems, ethernet connections between computers, or even data on a hard disk drive. There are several different methodologies to checking for data accuracy, but the main goal is to make sure that the data is correct.

The simplest way uses a parity bit. When writing data, the amount of non-zero bits are added up, and if the result is an even number, the parity bit is set to one. If the result is an odd number, the parity bit is set to zero.

Unfortunately, this method of adding up the bits does not ensure that we are always getting accurate data. In the event that an even number of bits are flipped, the resultant parity still passes, even though the data is incorrect.

The next step in data checking is to calculate a checksum for the data. An algorithm is defined as when the data byte or bytes is passed through it, a resultant number is calculated. This number is called the checksum. The algorithm is defined as achieving a different resultant checksum if the data has been corrupted during transmission. If defined properly, the algorithm's checksum, if matched, leads to a very high probability that the data is valid.

For example, a simple algorithm might be that the bytes in the transmission burst are added up, and then the last byte transmitted contains the checksum. The receiving controller then reads in all the bytes and adds up all the bytes except the last byte, which is then compared to the last byte to see if the data is valid.

The issue with this additive method is that there still exists instances where invalid data could still add up to the same checksum.

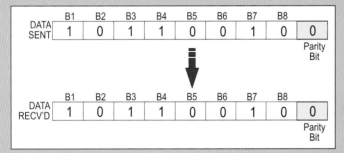

The data from the sender to the receiver is verified because the parity bit agrees between the two.

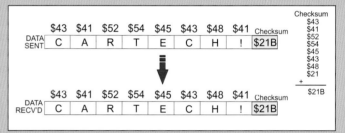

Here a 16-bit word is being passed between two nodes. The ASCII message passed from the sender to the receiver and was decoded into its hex equivalents. The value of each byte is then added up, and a checksum is determined.

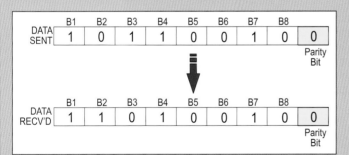

In this example, the data that the sender has sent is not equal to the data that was received. Unfortunately, the parity bit does not flag this error because an even number of bits are incorrect. This is the downfall of the parity checking scheme because it can only detect when an odd number of bits are corrupt.

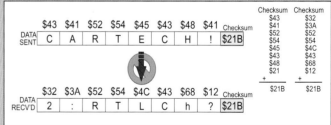

The data between the sender and receiver has become corrupt. Unfortunately, in this instance, even though the data is different, the same checksum was calculated and the receiver thought that the data was correct. Granted, the odds of this happening are very low, but it does illustrate that this particular error detection method is not guaranteed.

THE OBD-II DATA INTERFACE

Modern-day error checking usually involves cyclical redundancy checks (CRC). In fact, the SAE has its own eight-bit check aptly named SAE J1850 CRC-8. In this book, I concentrate on this particular CRC for examples, but there are many other CRC schemes, both published and not published.

The idea behind the CRC is to use polynomial math on the data. A polynomial is a mathematical equation constructed of both variables and constants containing only addition, subtraction, multiplication, and whole number, non-negative exponents.

A simple example of a polynomial is:

$$x2 + 4x + 11$$

where:
x = variable
4 and 11 = constants

Each CRC method has its own polynomial. In the SAE J1850 CRC-8 methodology, for example, the polynomial is:

$$x8 + x4 + x3 + x2 + 1$$

In a nutshell, and without going into an entire dissertation of the intricacies of CRC computation, the polynomial that is agreed upon by the sender and receiving nodes is used to divide the transmitted data and derive a checksum. The checksum between the sender and receiver is verified to determine if the data is valid.

---

connector. Referencing the table below and inspecting the pins on the J1962 connector may help you determine the protocol.

A simple check using the voltmeter can determine if the bus is ISO 15765. By setting the voltmeter to read resistance, and placing the red/positive lead on pin 6 and the black/negative lead on pin 14, the voltmeter should read a resistance of 60 ohms. If any other resistance is shown on the voltmeter, or the voltmeter registers the connection as open (0 ohms), then the vehicle does not communicate using ISO 15765.

Once the protocol is determined, the scan tool can be set to communicate using that protocol.

ECM Communications Hung Up

Occasionally, the ECM can get into a "hung" mode, in which it functions but does not communicate. The vehicle runs, but the scan tool reports that the communications link cannot be established. If all other checks have been verified, the ECM may need to be rebooted. To reboot the ECM, disconnect both battery leads and then press on the brake pedal to discharge any capacitors in the system. After the battery has been reconnected, the entire ECM reboots and should start to communicate again.

These two identical 2006 Chevrolet Corvettes have different software segments loaded. This discrepancy in segments may make it difficult for the scan tool to connect to this vehicle if it does not recognize the newer software versions.

CHAPTER 3

Before disconnecting the battery, consult the vehicle's owner's manual on the proper procedure.

Missing Data or Invalid Data

Most scan tools connect to the vehicle and query the ECM for valid PIDs for the vehicle. Some scan tools assume the last vehicle connected is the current vehicle. Thus, the PIDs may not match up. So when evaluating data, some PIDs may give improper data values and other PIDs may show up as not being available. To remedy this situation, scan tools allow you to query for the PIDs. This process may take a minute or more depending on the vehicle. When completed, the PID list should match the vehicle and the data should be accurate. Some scan tools may not automatically scan the PIDs, but they may rely on the vehicle's VIN to determine what PIDs are valid. As a general rule, I prefer to always scan for PIDs to make sure that the scan tool is reporting exactly what the ECM is providing. It is well worth the minute of query that it takes to ensure that the scan tool has proper data.

Unknown ECM Software

Much like the personal computer industry, the OEMs are constantly updating the software in the ECM. Many owners don't realize the number of technical service bulletins (TSBs) that exist for updates to the software of the ECM. When a vehicle is serviced at a dealership, if there is a TSB for upgrading the ECM's software, the dealership automatically updates the software and many times without the knowledge of the owner. This ensures that all the vehicle's controllers are operating at peak efficiencies at all times.

Unfortunately, some scan tools rely not only on the vehicle type, but also the software revision within the ECM and other controllers in order to download the proper PIDs and vehicle information. Thus, if a new version of the software for the ECM is available to the dealership and it was flashed into the ECM, the scan tool may not be able to properly identify the vehicle and may not connect to it.

This is why it is important when shopping for a scan tool to make sure that it can be updated. Many scan tools connect to a personal computer, and most of the updates are downloaded from reputable internet websites. Typically, an update to the scan tool's software or firmware is available on the scan tool manufacturer's website and available to the general public or to registered users. By keeping the scan tool up to date, it can recognize all the updated software revisions in the modules, as well as access new features available in the modules.

CHAPTER 4

Scan Tool Interfaces

The OBD-II system is capable of delivering a vast amount of data, including real-time operating conditions, diagnostic data, and bi-directional programming and testing. A variety of interfaces are available, ranging from a do-it-yourself electronic interface with simple RS232 terminal input; to an inexpensive generic scanner; all the way up to expensive, manufacturer-specific scan tools and programmers. The inexpensive scan tools typically allow the querying of generic OBD-II scan data. The higher-end scan tools allow querying of both generic scan data and manufacturer-specific scan data. Scan-tool selection depends on the depth of diagnostic capability that is needed or anticipated.

Generic OBD-II Scan Tool

SAE J1978 specification defines the generic OBD-II scan tool's required minimum functionality. This generic scan data information is common to all vehicles adhering to the OBD-II standards. This specification has the following requirements:

- Automatically determine the communication interface without any user interaction and connect to the system
- Display real-time data of vehicles sensors
- Query and display module, vehicle, and systems information
- Query and display the results of onboard diagnostics tests
- Display pending and current diagnostic trouble codes as well as freeze-frame data
- Clear any pending or current diagnostic trouble codes and erase freeze-frame data
- Provide limited help on information to the end user

The benefit of the generic scan-tool data is that a single scan-tool interface can be used to diagnose a variety of model years and manufacturers. The mechanic is able to diagnose and correct the problem on a variety of vehicles from most of the data available from the scan tool interface. Therefore, this is a very popular level for home enthusiasts, due to its inexpensive cost.

Manufacturer-Specific OBD-II Scan Tool

Vehicle systems have become more complex, and manufacturers have implemented more control modules to query and manage a variety of aspects of the vehicle's systems. As a result, some control modules do not provide generic OBD-II scan-data specifications, but the scan tool needs to poll information from these modules. The OBD-II specification allows an extended range of OBD-II codes that the manufacturer can assign queries for specific control modules, sensors, and entire subsystems.

Unfortunately, as was witnessed in the OBD-I days, confusion can exist as manufacturers' implementations can overlap diagnostic information codes. They can choose whether or not to display real-time data and they can dictate the format of the data. Furthermore, some manufacturers have vehicle diagnostic information that may exist outside the realm of OBD-II scan tools, such as vehicle navigation systems, multimedia equipment status, etc. Also, the manufacturer-specific toolsets can allow further diagnostics, such as the capability to cycle solenoids, turn components on and off, and exercise emissions systems.

All of this helps diagnose specific issues. For example, the SAE specifications require the final ignition timing to be reported in terms of degrees. Unfortunately, this doesn't show the entire picture of what is occurring. In a General Motors–based vehicle, manufacturer-specific data fields can be queried to fully

AUTOMOTIVE DIAGNOSTIC SYSTEMS

CHAPTER 4

understand what is occurring in terms of ignition timing, including knock retard values, spark smoothing value, engine-coolant-temperature spark adjust, intake-air-temperature spark adjust, adaptive spark (octane modifier), and catalyst-light-off spark. Thus, the specific manufacturer's information can give a more detailed picture of what is occurring than the generic, required OBD-II scan data.

Reading Scan Data

Some scan tools do not differentiate between generic and manufacturer-specific scan data and do allow them to be queried simultaneously. Other scan tools require the mechanic to choose which type of scan data is being queried or what manufacturer made the vehicle in order to show the specific data applicable to the vehicle. Typically, when diagnosing a vehicle, it is very convenient to show generic and manufacturer-specific data at the same time. The manufacturer-specific data typically supplements the generic data when trying to get a good understanding of the current status of the vehicle, especially when looking at real-time data. DTCs are formatted so that the mechanic can tell if the trouble code is a generic trouble code or a manufacturer's. The scan tool interface may or may not differentiate between the two formats, though; so, when in doubt, consult the service manual. (See Chapter 6 for details on this format.)

Additional scan tool data can come from controller modules other than the ECM. As the complexity of vehicles increases, more electronic control modules that control complex systems are being added, such as active handling, passive restraint systems, climate systems, etc. These modules have their own diagnostic capabilities and data scanning capabilities that may or may not be separate from the ECM. So the manufacturer has expanded the scan-data toolset to include specific information from the other modules.

Obviously, these control modules vary among vehicles and manufacturers, and aren't covered by the generic scan-data protocols. Therefore, when diagnosing these systems, a manufacturer-specific scan tool is often required, or a robust scan tool that supports the manufacturer-specific data, as well as technical documentation on the various control modules. Some scan tools contain a data dictionary for manufacturer-specific code, so consult the scan tool manual for details.

The issue with scan tool data is whether or not the interface supports some or all of the manufacturer-specific data sets. Unfortunately, many of the scan tools do not support the entire set of manufacturer-specific data. Also there are very few scan tools, except for the very-high-end scan tools, that support the entire set of every manufacturer's specific codes. Because the manufacturers are constantly modifying these data sets, some scan tool vendors have online update capabilities to allow mechanics to keep up with the changing technology. These updates/upgrades can be free or they may be available on a monthly/yearly paid subscription service.

OBD-II Trouble Code Reader

A scan-tool-interface subset of the generic OBD-II scan tool is the trouble code reader scan-tool interface. Many of these scan tools are available at auto parts stores. The main purpose of these "light" scan tools is to query the diagnostic trouble codes stored in the ECM and clear them as necessary. These are the modern-day version of the OBD-I paper clip for retrieving "blinky codes." They are convenient—able to query what the code is and clear it—but, without any real-time data, they don't easily solve most of the problems that require diagnosis. Luckily, the cost of these scan tools is very low, so it offers an entry into OBD-II diagnostics for very little cost. Because they are also relatively compact, they are handy to have in the glove box to quickly look at codes and clear them without having to pull out a larger, high-end scan tool. This is especially true if the scan tool is laptop based.

Benefits of a trouble code reader:

- Read current diagnostic trouble codes to aid in the identification of the cause of a set OBD-II diagnostic alert
- Read pending diagnostic trouble

On certain scan tools, the mechanic has to choose what type of scan data to review.

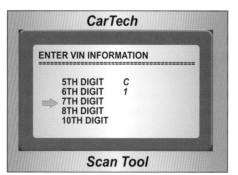

When a scanner is connected to the system, some ECMs do not correctly identify the make and model of the vehicle. In these cases, the mechanic needs to enter the VIN into the scanner, so that manufacturer-specific codes and the codes available for the particular vehicle can be determined.

codes to aid in the identification of the cause of a continuous OBD-II diagnostic alert
- Determine and clear the status of the MIL
- View the state of the Inspection/Maintenance (IM) monitors
- Inexpensive, many retailing for less than $35
- Very simple to use
- Self powered using 12-volt power source from J1962 connector

- Requires little knowledge of OBD-II diagnostics

Possible disadvantages of a trouble code reader:

- Limited or no manufacturer-specific codes supported
- Only points to diagnostic trouble codes that are set without any more information for diagnosing what caused the error code

- Scan tool software may not be able to be updated as new model years come out
- Freeze-frame data may not be available
- Real-time run data may not be available or limited data tracing
- Minimal or no help files stored on device for codes and IM monitors
- May not be compatible with European-manufactured automobiles

Performing Simple Diagnostic Procedure

1 Diagnostic Trouble Code Reader

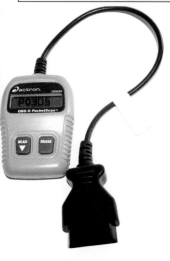

The basic tool for reading trouble codes is small, inexpensive, and easy to use.

2 Plug the DTC into the J1962 Connector

Find the J1962 connector. It should be located within 3 feet of the steering wheel. Plug in the cable from the trouble code reader to the J1962 connector. It is keyed to eliminate plugging it in incorrectly.

3 Place Ignition Switch in the Run Position

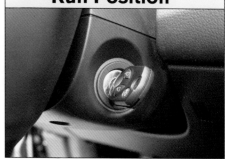

The ignition key must be in the run mode in order for communications to take place. The engine may or may not be running. Reading trouble codes does not require the engine to be running.

4 Read the Scan Tool Data

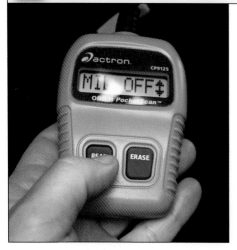

On this particular model of reader, the "READ" button is pressed to scroll through a variety of trouble codes, misfires, monitors, and other information. I recommend having a pencil and paper to write down any of the codes that appear.

5 Erase Any DTC Codes

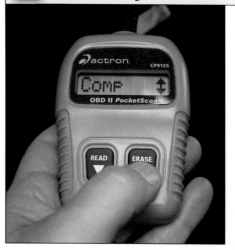

To erase any trouble codes that are set, this model of reader has an "ERASE" button, which is pressed and held to erase the codes. Keep in mind that erasing any codes typically resets all system readiness monitors.

CHAPTER 4

Entry-Level OBD-II Scan Tool

The enthusiast-level of OBD-II scan tool is targeted toward the home mechanic. This scan tool represents a step up from the basic trouble code reader because it allows investigative diagnostic information to be queried from the ECM. This information then can be used to map out a strategy to diagnose and correct the issue. These scan tools vary in price from less than $100 to around $500. They are available at most automotive parts stores as well as online.

Benefits of an enthusiast-level scan tool:

- Read current DTCs to aid in the identification of the cause of a set OBD-II diagnostic alert
- Read pending DTCs to aid in the identification of the cause of a continuous OBD-II diagnostic alert

Choosing a Scan Tool

Choosing a scan tool can be confusing as there are a myriad of choices available, ranging from the inexpensive, simple trouble code reader to a complete, stand-alone diagnostic unit that can cost tens of thousands of dollars. The decision comes down to a few key factors: capabilities required of the scan tool to determine how in-depth the repairs are going to be, how many different vehicles are going to be diagnosed, and how realistic the budget is.

Capabilities Required

The capabilities of the scan tool really depend on the information that is going to be required to perform the diagnostics, and your comfort level with the repairs to be performed. An inexpensive, simple trouble code reader shows the diagnostic trouble codes that are set and can lead you to the possible area of the issue. Unfortunately, this really only provides the information from a very generic level. Typically, further diagnostics are required to understand what caused the trouble code to be set in the first place. Trouble code readers are handy for seeing what DTCs are set, but may not accurately diagnose the cause of the issue.

To accurately diagnose an issue or symptom, more data must be gleaned from the system. Generic, capable scan tools handle most of the trouble codes and diagnostic capabilities to repair most common sensor issues, misfire issues, and performance issues. These repairs are easily done in the home garage and can be resolved with parts available at most local auto parts stores. Thus, this type of scanner should work for roughly 90 percent of the issues that occur.

When troubles occur in manufacturer-specific modules and capabilities, professional-level scan tools are typically required. These are needed to diagnose issues in control modules such as anti-lock braking systems, drive-by-wire modules, climate modules, etc. These advanced capabilities in the professional scan tool are a must when trying to repair these systems. Unfortunately, due to the complexity of many of these systems, and the lack of replacement parts at your local auto parts store, many of these repairs are best left to professional repair shops.

Vehicles to Diagnose

When choosing a scan tool, it is critical that it not only support all of the generic PIDs, but it should also support the manufacturer-specific PIDs for vehicles that are to be diagnosed. It is critical to be able to support this extended manufacturer data to be able to accurately pinpoint any issues.

Budget

Scan tools (along with trouble code readers) can vary in price from as inexpensive as $30 all the way to exceeding $10,000. Many auto parts stores scan codes for free, which can be a good starting point when beginning to diagnose OBD-II issues. Budget typically drives capabilities, but at a minimum, to accurately and quickly diagnose problems, the following features should be considered in a scan tool:
- Supports generic and manufacturer-specific PIDS
- Reads and clears diagnostic trouble codes
- Reads freeze-frame data
- Logs real-time data
- Queries status of system monitors
- Offers online help for DTCs and PIDs
- Includes upgradable software for new makes, models, and years
- Has large display and is easy to read in daylight conditions

SCAN TOOL INTERFACES

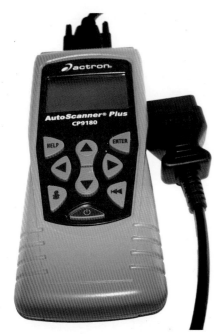

The entry-level scan tool gives more diagnostic information than just reading diagnostic trouble codes. Many of these scan tools can display real-time data information, which is useful when diagnosing sensors and feedback/control systems. Basic information, such as fluid temperatures, air temperatures, timing, fuel control, and engine speed, are considered the basic requirements for an enthusiast's scan tool.

- Determine and clear the status of the MIL
- View the state of the IM monitors
- Supports all generic and most domestic manufacturer-specific OBD-II trouble codes
- View and log real-time performance data
- Query six of the ten available OBD-II modes
- Determine vehicle VIN, CVN, and calibration ID of on-board software
- Self powered using 12-volt power source from J1962 connector
- Hand-held size for easy portability
- Most allow some sort of free upgrade from downloading over the internet, or requesting a CD-ROM of new software from the manufacturer.

Disadvantages of an entry-level scan tool:

- May require additionally purchased adapters and cables to work with different vehicles
- Updates to the software require a connection to a computer and internet access
- Cost may be prohibitive for occasional use
- Small screens may be difficult to view large quantities of real-time data
- Limited diagnostic information and onboard help
- Limited run-time data-save size
- May require a battery to save run-time data and a PC computer to review captured data

Professional-Level OBD-II Scan Tool

The professional OBD-II scan tools are targeted toward repair shops, which require scan data and diagnostic data from a variety of automobile makes and models. Often the previous generation of scan tools are available at relatively low cost as repair shops upgrade their tools, putting these tools within reach of home enthusiasts.

Benefits of a professional-level scan tool:

- Read current DTCs to aid in the identification of the cause of a set OBD-I or OBD-II diagnostic alert
- Read pending DTCs to aid in the identification of the cause of a continuous OBD-II diagnostic alert
- Determine and clear the status of the MIL
- View the state of the IM monitors
- Supports all generic and all manufacturer's specific OBD-I and OBD-II trouble codes
- View and log real-time performance data using tabular and graphical views
- Query all ten available OBD-II modes and communicate with other vehicle-based control modules
- Determine vehicle VIN, CVN, and calibration ID of onboard software
- Self powered using 12-v power source from J1962 connector

The Snap-On Tools Solus Pro Scanner kit includes a universal SAE J1962–based cable and personality keys along with a charger and case. Documentation is typically included on multiple CDs and is also installed on the scan tool itself, eliminating the requirement to reference hard documentation when looking up codes and other information.

AUTOMOTIVE DIAGNOSTIC SYSTEMS

- Data saved on a variety of removable RAM devices (flash drives, stick memory, etc.) that can be read back later or transferred to a standard PC
- Software can be updated via an internet connection
- Most are based on PC Windows operating systems built into the unit
- Built-in diagnostic help menus and sensor information
- Additional diagnostic tools, such as oscilloscope and voltage meters, are built-in
- Capable of testing individual sensors

Disadvantages of a professional-level scan tool:

- Prices start at several thousand dollars and go up from there
- Older-version scan tools require multiple personality keys
- Must have on-board batteries charged in order to operate
- Some scan tools require a monthly/yearly subscription to be purchased to allow upgrades/updates

Personality Keys and Adapters

Personality keys were developed to connect to a variety of vehicles without having to purchase a multitude of different cables. Based on the VIN or entered information, the scan tool tells you which personality key is required to interface with the vehicle. The scan tool typically has a variety of different inserts that fit into the J1962 connector, which configure it to the particular vehicle. Recall that even though the J1962 connector is standard, the pin configurations based on one of the communications protocols are different. The personality key establishes which pins are used for what information via the mechanical connection that it makes.

Manufacturer-Specific OBD-II Scan Tool

Many automotive manufacturers develop a scan tool specifically targeted toward their vehicles. Most of these are located at the automotive dealerships where the majority of repairs are performed on their makes of vehicles. Although they do support the entire suite of generic codes, they only support the specific manufacturer's codes of their own brand of automobile.

One advantage of these scan tools is that they also have the capability to completely rewrite the ECM's on-board, flash-based RAM. It can repair a damaged ECM as well as update the automobile's software due to recalls, upgrades, etc. Most manufacturers have dedicated control units for non-engine-related control (i.e., multimedia systems, traction control systems, drive-by-wire, etc.), and standard scan tools do not interface with these systems. Often the manufacturer-specific scan tool must be used to communicate with these modules and to diagnose these non-drivetrain systems.

Personal-Computer-Based OBD-II Scan Tool

The portability of the personal computer, the open specifications of OBD-II communications protocols, and the ease of software development have

Personality keys configure the scan tool to a particular vehicle's communications protocol. Additional keys can be purchased as necessary.

Here, the proper personality key K-5A is installed in the slot of the cable, and it allows the scan tool to communicate with this particular vehicle. Unfortunately, the modular design of the cables allows a variety of different connector shapes that may not conform to the SAE J1962 standardization to be used by the scan tool.

led to an entire industry of scan tools that run on personal computers. These scan tools range from simple code readers, with minimal real-time data scanning, to complete scan tools with flash programming capabilities. Typically, these systems include a cable and a CD-ROM containing the software, which is loaded on a laptop or other personal computers. With the increase in the power of hand-held computers, new scan tools are available on palm-size computers and personal assistants. Use care when choosing a personal computer-based scan tool; their capabilities vary considerably. Also, most personal computer-based scan tools are limited as to which vehicle manufacturers that they can scan.

Because the standards for communications protocols are rigidly defined, many third-party hardware manufacturers have developed communication cables that connect the personal computer to the OBD-II interface. ELM Electronics offers an innovative and inexpensive chip that converts the different OBD-II protocols to the standard RS-232 protocol (which personal computers use to communicate). Thanks to this inexpensive, easy-to-develop chip set, a lot of inexpensive OBD-II interfaces have been introduced to the market under the ELM name. The software is offered from no-cost freeware to professionally developed software, which can be very expensive. Most of the software is available on a trial basis so, after purchasing the ELM-based interface, you can test a variety of software to see if it has the capabilities that are required.

Some companies have developed both hardware and software that function as a scan tool running on a personal computer. This software starts on the low end as unsupported freeware and goes up to professional-level software, which can cost several hundred dollars. When choosing software, look for the software with the capabilities for the work you are performing, and from a company that provides adequate customer support. Also the company should continue to improve functionality and offer software support for little or no additional cost.

Other companies have produced proprietary cables and software sold as a set. An Internet search of OBD-II scan tools yields a variety of choices. Finally, if you're in a pinch and if your need is specific, many auto parts stores can scan a vehicle and print out a report for little to no charge.

An example of a manufacturer-specific scan tool; in this case, a General Motors' Tech II scan tool kit.

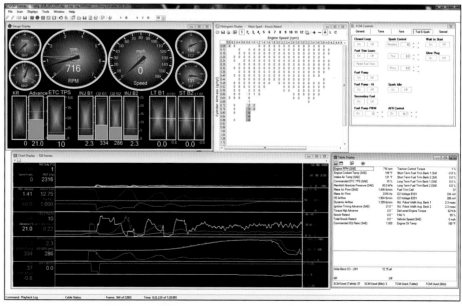

A screen shot of HP Tuners' VCM Scanner scan tool running on a laptop personal computer. (Screen courtesy HP Tuners)

CHAPTER 5

OBD-II MODES

The OBD-II system has many diagnostic modes to identify and troubleshoot problems within the system. The ISO 15031 specification is based on the open systems interconnection (OSI) model, which is defined by ISO/IEC (International Electrotechnical Commission) 7498 and ISO/IEC 10731. The foundation of the OSI model divides network communications into seven distinct layers:

- Physical layer
- Data link layer
- Network layer
- Transport layer
- Session layer
- Presentation layer
- Application layer

The OSI model is generically defined and not intended to specify particulars such as communications protocols, programming languages, computer hardware, etc. The emissions-related standards fall into some of those seven defined layers.

The ISO 15031 specification defines the communications protocol that the OBD-II uses, and it falls under the application layer defined by the OSI model. The ISO 15031 is used across all the various OBD protocols, and therefore has become a standard of scan-tool definitions. The ISO 15031 specification defines the OBD communications protocol, which is used in nine distinct modes. A unique hexadecimal number from $01 to $09 defines each of these modes. (The "$" is standard notation to alert readers and programmers that the number displayed is a base-16 hexadecimal number and not a number from another base. This annotation is followed throughout this book.)

Each one of the nine modes conveys specific types of diagnostic information that is important when diagnosing a vehicle issue. Professional-level scan tools typically have the capability to query all of the nine modes, but enthusiast-level scan tools may only be able to access a few of them. When choosing a scan tool, be sure that the required number of useful modes (including $01, $02, and $03) are available for diagnostic duties. The additional modes can provide more information about the status and diagnostic information for the vehicle.

Understanding what information each mode can provide helps to determine if a particular scan tool will help or hinder diagnostics.

A scan tool considered to be completely "OBD-II-compliant" should be able to access all nine modes. Some compliant scanners have poor coverage

Applicability	OSI Model Layer	Emissions Related Diagnostic Standards			
Seven layer according to ISO/IEC 7498 & ISO/IEC 10731	Physical (1)	ISO 9141-2	ISO 14230-1	SAE J1850	ISO 11898 ISO 15765-4
	Data Link (2)	ISO 9141-2	ISO 14230-2	SAE J1850	ISO 11898 ISO 15765-4
	Network (3)	—	—	—	ISO 15765-2 ISO 15765-4
	Transport (4)	—	—	—	—
	Session (5)	—	—	—	ISO 15765-4
	Presentation (6)	—	—	—	—
	Application (7)	ISO 15031-5	ISO 15031-5	ISO 15031-5	ISO 15031-5

This is the standard OSI layer model and how the OBD-II diagnostic specifications overlay it.

Mode $01 - Request Data by Specific PID

Mode $01 is also known as the generic OBD-II mode, because it is the most common mode used to gather information. Most basic scan tool devices have access to some or all of the information in this mode. Trouble code readers typically do not include information from this mode. Mode $01 contains the following information sets:

- Available sensor PIDs
- OBD-II self-check monitors, continuous and non-continuous
- Calculated performance values

PID 00

Mode $01 PIDs contain the majority of common diagnostic information used. In fact, PIDs (parameter IDs) are used within OBD communication. These IDs are expressed as codes defined by SAE J1979. The codes are used to request and return applicable diagnostic information. Although the specification contains a variety of PIDs, the vehicle in question may not use all of them. So when a scan tool is connected, it typically polls the ECM to determine what PIDs are available. A query of the ECM returns a bit stream of information to determine which of the PIDs are valid for the connected ECM. After the available PIDs have been determined from the initial scan, the PIDs are then queried to ensure that they are reporting the proper information. This PID count varies from vehicle type to vehicle type, as well as between different manufacturers. Therefore, the scan tools always query and verify PIDs prior to reporting their data.

OSI Model Layers

The structure of the OSI seven-layer model was developed in the late 1970s, and it was based on the development of the first computer networks. The goal of the specification was to divide the network into specific layers that interconnected or shared information only with the layers immediately above or below the particular layer. The OSI model is adaptable on a wide variety of network communications, such as the diagnostic network in a vehicle or a data network in a corporate computer facility.

Each layer defines a particular specification in the network. The layers are defined as follows:

Application layer: Interacts with the software, and it's essentially the end user who interacts with this layer. In terms of OBD, ISO 15031 defines all of the modes. The user interacts with the ECM and various modules to query and receive data about the vehicle and DTCs.

Data layer: The procedures used to transfer the data between the physical devices and how to detect errors within the data transfer. In emissions terms, many specifications overlap with the physical layer definitions or are subsets of them, and these define the error handling between the devices. Vehicle networks often use CRC or cyclic redundancy checks.

Network layer: How to handle variable length data, which is transmitted. In an automobile, only the CAN network handles the variable-length data packet, and therefore it is the only protocol that can be mapped to the network layer. All the other OBD protocols use a fixed-length data packet.

Physical layer: The physical hardware used between network devices (wire type, connectors, adapters, pinouts, etc.) as well as the electrical properties of the devices (voltages, logic levels, etc.). In terms of OBD specifications, this is where the connector types are defined, along with pinouts, voltage levels, protocols, etc.

Presentation layer: Can be thought of as an interpreter between the higher-functioning applications and the lower-level communication. It is the go between that handles all of the translations, and thus it frees the applications from having to do so. There is no corresponding specification of OBD protocols.

Session layer: Manages the connection between hardware nodes. In the event of a failure, it is also responsible for terminating and rebooting modules on the network. Only the CAN bus (ISO 15765) provides this type of management between control modules.

Transport layer: Responsible for providing the data transfer service to the upper layers. It is also responsible for the handshaking for the data, and that means it ensures that the data arrives correctly and waits until the receiver is ready to accept more data. There is no equivalent to this layer in OBD protocols.

CHAPTER 5

Certain scan tools store the available PIDs by the VIN. As a result, the PID scan does not have to be scanned each time the scan tool is connected to the vehicle. This feature can save time if a particular vehicle is scanned several times because certain vehicles can take several minutes to complete the initial query of PIDs. Newer vehicles also poll other control modules that are located on the system bus. These can include body control modules, active handling modules, transmission control modules, etc.

During the initial connection, the scan tool polls PID 00 to determine how many PIDs are available and which ones they are. Depending on the vehicle make and model, this process can take from 30 seconds to several minutes.

PID $01

PID $01 returns to the scan tool a 4-byte string of data that contains the current status of the diagnostic codes. It returns the current status of the MIL, whether it is currently lit or not. It also returns the status of the number of diagnostic trouble codes currently set within the ECM. The following continuous diagnostic tests are represented:

- Misfire
- Fuel system
- Component tests

There are also allowances reserved for additional diagnostic tests that the manufacturer can opt to include. In addition to the continuous diagnostic tests, the following non-continuous monitoring tests are reported:

- Catalyst
- Heated catalyst
- Evaporative system
- Secondary air system
- Air conditioning refrigerant
- Oxygen sensor
- Oxygen sensor heater
- Exhaust gas recirculation (EGR) system

PID $02

PID $02 does not exist in Mode $01. It only exists in Mode $02.

PID $03

PID $03 is unique in that it returns an encoded status of the fuel system. Possible values of this PID can be:

- Open loop due to insufficient engine temperature
- Closed loop full, sensor feedback
- Open loop due to engine load
- Open loop due to sensor failure
- Closed loop, but one oxygen sensor is faulted

This is a key PID to watch because it tells if the variety of sensors is being used to determine fueling of the vehicle. This key PID also determines if there has been a sudden loss in performance or fuel mileage or if there is a suspected problem with a fuel-system feedback-loop component.

PID $04 to $4E

PIDs $04 through $4E return a variety of information about the conditions current at the time of the scan. Note that not all PIDs are supported on all vehicles.

What Modes Are Commonly Used?

The more modes a scan tool supports, the more information available to diagnose a problem. The first mode that must be used when using the scan tool is Mode $03. This shows the currently set diagnostic trouble codes.

Once the codes are determined, Mode $02 is typically accessed next to show what the running conditions were when the fault was found.

To determine if the fault is still ongoing, Mode $01 is accessed to look at the real-time data and the values are examined to further help diagnose the cause of the fault.

Once the determination is made as to what to replace and the repair is made, Mode $04 is used to clear the faults.

Finally, Mode $01 is used again to confirm that the sensor or repair just completed is functioning properly.

On scan tools that support Mode $06, it is helpful to use Mode $06 to look at the monitors testing in real time. This helps validate the repair just made, or show other issues that are occurring. Remember, a MIL is illuminated after an error has occurred. By using Mode $06, problems can be witnessed and diagnosed well before they reach the level of criticality where they trip the diagnostic trouble code.

OBD-II MODES

Specification SAE J2190 defines these PIDs, which are mapped to the following information with their units shown in parentheses if applicable:

PID	Description
$04	Calculated engine load value (percent)
$05	Engine coolant temperature (°C)
$06	Short-term fuel trim (STFT) Bank 1
$07	Long-term fuel trim (LTFT) Bank 1 (percent)
$08	Short-term fuel trim (STFT) Bank 2 (percent)
$09	Short-term fuel trim (STFT) Bank 2 (percent)
$0A	Fuel Pressure (kPa)
$0B	Intake manifold pressure (kPa)
$0C	Engine speed (RPM)
$0D	Vehicle speed (km/hour)
$0E	Timing advance (degrees)
$0F	Intake air temperature (°C)
$10	Mass Air Flow rate (grams/sec)
$11	Throttle position (percent)
$12	Secondary air status
$13	Oxygen sensors (O_2) present
$14	O_2 Bank 1 Sensor 1 (voltage)
$15	O_2 Bank 1 Sensor 2 (voltage)
$16	O_2 Bank 1 Sensor 3 (voltage)
$17	O_2 Bank 1 Sensor 4 (voltage)
$18	O_2 Bank 2 Sensor 1 (voltage)
$19	O_2 Bank 2 Sensor 2 (voltage)
$1A	O_2 Bank 2 Sensor 3 (voltage)
$1B	O_2 Bank 2 Sensor 4 (voltage)
$1C	OBD-II standard definition
$1D	Oxygen sensors (O_2) present
$1E	Auxiliary system status
$1F	Run time since engine start (seconds)

Additional PIDs between $21 and $40 may be used on a vehicle. PID $20 can be queried and will return the PIDs that are valid for the ECM. SAE J2190 defines these PIDs.

PID	Description
$21	Distance traveled since MIL (km)
$22	Fuel rail pressure ref. to MAP (kPa)
$23	Fuel rail pressure for diesel (kPa)
$24	O_2 Sensor 1 lambda equivalence (V)
$25	O_2 Sensor 2 lambda equivalence (V)
$26	O_2 Sensor 3 lambda equivalence (V)
$27	O_2 Sensor 4 lambda equivalence (V)
$28	O_2 Sensor 5 lambda equivalence (V)
$29	O_2 Sensor 6 lambda equivalence (V)
$2A	O_2 Sensor 7 lambda equivalence (V)
$2B	O_2 Sensor 8 lambda equivalence (V)
$2C	EGR commanded (percent)
$2D	EGR error (percent)
$2E	Evap purge commanded (percent)
$2F	Fuel level (percent)
$30	Number of warm ups since DTC cleared
$31	Distance traveled since DTC cleared (km)
$32	EVAP vapor pressure (Pa)
$33	Barometric pressure (kPa)
$34	O_2 Sensor 1 lambda equivalence (mA)
$35	O_2 Sensor 2 lambda equivalence (mA)
$36	O_2 Sensor 3 lambda equivalence (mA)
$37	O_2 Sensor 4 lambda equivalence (mA)
$38	O_2 Sensor 5 lambda equivalence (mA)
$39	O_2 Sensor 6 lambda equivalence (mA)
$3A	O_2 Sensor 7 lambda equivalence (mA)
$3B	O_2 Sensor 8 lambda equivalence (mA)
$3C	Catalyst Bank 1 Sensor 1 temp (°C)
$3D	Catalyst Bank 2 Sensor 1 temp (°C)
$3E	Catalyst Bank 1 Sensor 2 temp (°C)
$3F	Catalyst Bank 2 Sensor 2 temp (°C)

Additional PIDs between $41 and $60 may be used on a vehicle. PID $40 can be queried and will return the PIDs that are valid for the ECM. SAE J2190 defines these PIDs, and some may be reserved for future usage.

PID	Description
$41	Monitor status drive cycle
$42	Control module voltage (V)
$43	Absolute load (percent)
$44	Commanded equivalency ratio

Freeze-frame data shows the exact conditions when the diagnostic trouble code P0301 (cylinder number-1 misfire) was set.

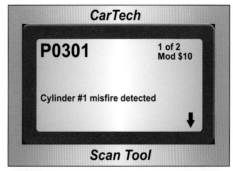

The scan tool requested and received the set diagnostic trouble codes indicating that there are two DTCs currently set. The first one showing P0301 is a misfire on cylinder number-1.

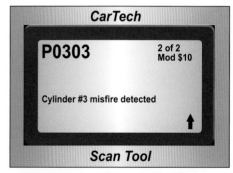

Most scan tools allow scrolling through the set DTCs by use of an arrow key or scroll pad. Here, the second set DTC, a misfire on cylinder number-3 (P0303) is shown as the second of two set DTCs.

CHAPTER 5

$45 Relative throttle position (percent)
$46 Ambient air temperature (°C)
$47 Absolute throttle position B (percent)
$48 Absolute throttle position C (percent)
$49 Absolute throttle position D (percent)
$4A Absolute throttle position E (percent)
$4B Absolute throttle position F (percent)
$4C Commanded throttle actuator (percent)
$4D Time engine running since MIL (minutes)
$4E Time since DTC cleared (minutes)

A complete description as well as values for the more common PIDs is addressed in the sidebar "Commonly Referenced PIDs."

Mode $02 - Request Freeze-Frame Data by Specific PID

Mode $02 is used to help diagnose trouble codes that are currently set. This mode returns the freeze-frame data, which reports the exact conditions when the fault was set. Mode $02 contains the same PIDs that Mode $01 contains from PID $00 through PID $0D. Mode $02 also contains PID $02, which returns the specific trouble code that caused the freeze-frame data to be stored. Multiple codes may be stored, as well as multiple instances of freeze-frame data.

Mode $03 - Request Set Diagnostic Trouble Codes

Mode $03 returns the stored 5-digit diagnostic trouble codes that have been set. This mode makes the P0 trouble codes universally accessible by scan tools and trouble code readers. Higher-end scan tools may be required to read P1-level

Commonly Referenced PIDs

PIDs are the variables that hold the data that lets you record running conditions of the vehicle being diagnosed. PIDs fall into three categories:

Calculated PIDs: Calculating its value based on sensor readings and/or table lookups provides this data. There are standard calculations that all manufacturers use as well as manufacturer-specific calculations. For instance, the PID that measures the air mass per cylinder is a calculation of the value read from the mass airflow meter, multiplied by 15, and then divided by the engine speed in RPM.

Manufacturer-specific PIDs: These data points are calculated or data read from sensors. This data is usually specifically targeted to a manufacturer's implementation of equipment, or features that are unique to its vehicles. Support for these PIDs is dependent on the capabilities of the scan tool.

SAE PIDs: These are standard sensors or readings common to all vehicles regardless of the manufacturer.

The following information can be used to better understand what each PID is trying to convey. Only the more commonly used PIDs are covered here. For PIDs that are not covered here, consult a service manual or search online for more information.

Calculated Engine Load Value: Measured in terms of percentage, it varies from a 0-percent to 100-percent value. This calculation is used for a variety of settings including establishing ignition timing, amount of fuel to inject into the engine, transmission shift patterns, and so forth. Typically, this calculation is based on current airflow from the mass airflow meter divided by peak airflow. Simply put, the higher the percentage, the harder the engine is working. Vacuum leaks, dirty mass airflow meters, and incorrect manifold air pressure sensors can cause this reading to be incorrect.

Commanded Equivalency Ratio: Measured in a ratio. This is the commanded air/fuel ratio that the ECM is driving the engine toward. Typically, once the engine is warmed, this equivalency ratio should be near stoichiometric (14.6:1).

Engine Coolant Temperature (ECT): Measured in degrees Celsius. This is a direct measurement of the temperature of the engine coolant in the engine block or cylinder head.

Engine Speed: Measured in revolutions per minute (RPM). Occasionally, this is expressed in the SI units of Hertz (Hz). (Hertz can be converted by dividing RPM by 60.) The engine speed is measured by a tachometer signal generated by the electronic ignition or by a crank position sensor located in the engine block.

OBD-II MODES

trouble codes, as well as additional codes generated by other control modules (ABS systems, body control module, media control, etc.).

Mode $04 - Clear Stored Diagnostic Trouble Codes and Reset MIL

Mode $04 is used to clear any stored trouble codes and also reset the OBD-II system readiness monitors. Most scan tools and trouble code readers can access this mode. Some advanced scan tools may be required to clear codes set by additional control modules.

Mode $05 - Oxygen Sensor Test Results

Mode $05 is useful in the diagnosis of trouble codes possibly set due to malfunctions with the oxygen sensors. The scan tool is used to monitor the voltages reported by the oxygen sensors, including their minimum and maximum voltage values. Many newer vehicles do not support this legacy test.

Mode $06 - Advanced Diagnostic Mode

Mode $06 is only available on full-feature scan tools. It is jokingly referred to as the "missing mode" because it is usually hard to find on most scan tools, and even harder to interpret its returned data from the ECM. Actual test results for non-continuous monitors, such as catalyst monitors, O_2 sensor monitors, and EVAP system monitors are returned as hex values to the scan tools. With the advent of the controller area network (CAN), misfires are also monitored in Mode $06. Please note that on some manufacturers' ECMs, all Mode $06 data is erased and reset when the ignition is turned off.

The scan tool may automatically translate the returned hex codes and data

Intake Air Temperature (IAT): Measured in degrees Celsius. This measures the temperature of the incoming airflow into the engine. The sensor should read near the ambient air temperature on a cold, non-running vehicle.

Intake Manifold Air Pressure (MAP): Measured in kilo Pascal (kPa). As the engine runs, vacuum or negative pressure builds in the intake manifold. This PID measures that pressure, and on a normally aspirated engine, it reads between 0 kPa and 100 kPa. Vehicles with superchargers or turbochargers read above 100 kPa when they enter boost. When the engine is off, the MAP sensor should read 100 kPa or very near to it.

Long-Term Fuel Trim: Measured in percentages. This is a historical correction to fueling based on the closed-loop feedback fueling system. Values typically range from −25 percent to +25 percent, although some vehicles can go quite a bit higher.

Mass Airflow Rate: Measured in grams per second. The mass airflow meter reports how much air is coming into the engine. This mass airflow meter may actually return values as a voltage or a frequency, depending on the vehicle.

O_2 *Bank X Sensor X:* Measured in voltage. The oxygen sensors return the readings of fuel ratios in terms of voltage. The readings range between 0 volts and 1 volt. The smaller the voltage, the leaner the air/fuel ratio is.

Short-Term Fuel Trim: Measured of percentages. This is an instantaneous correction to fueling based on the closed-loop feedback fueling system. Values typically range from −25 percent to +25 percent, although some can go quite a bit higher.

Throttle Position Sensor (TPS): Measured in percentages. The TPS reading directly correlates to accelerator pedal position. Readings can range from 0 to 100 percent, with larger numbers indicating a more open throttle. On vehicles equipped with drive-by-wire systems, the TPS reading may never actually go to 0 percent.

Timing Advance: Measured in degrees. Typically, timing is expressed in terms of after top dead center (ATDC) or before bottom dead center (BBDC). Values can vary from negative to positive. Timing advance is expressed in terms of degrees, depending on where the ignition system is firing the spark plug in relation to the piston position.

Vehicle Speed: Measured in kilometers per hour or in miles per hour. This is typically read from a vehicle speed sensor (VSS) located in the transmission or in the wheel. Drive gear ratios and non-standard tire sizes can affect the accuracy of this reading.

CHAPTER 5

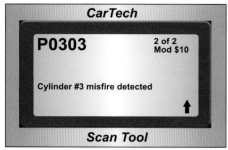

Mode $06 monitor tests shown for this vehicle which consists of tests $01, $03, $10, $26, and $27. Depending on the make and model of the vehicle, a variety of different tests can be performed in Mode $06. When selected on the scan tool, the supported tests will show allowing the technician to select the appropriate test.

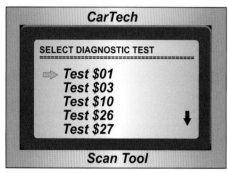

This shows returned data from Test $01 in Mode $06. You can cross reference this data in technical repair manuals or online at sites such as www.iatn.net, and you should discover what monitor this is testing and what the acceptable values are.

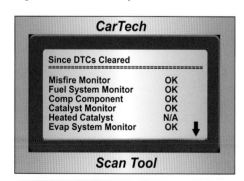

The scan tool shows the current status of the continuously monitored tests. On some vehicles, not all monitors are supported and are indicated as such. (Screen courtesy HP Tuners)

returned from the Mode $06 test monitors. If it does not, there are a variety of resources that can decipher the code, ranging from technical repair manuals and OEM websites to technical websites. Unfortunately, the OBD-II standards do not apply in Mode $06. Each vehicle may use the same test number to report on different monitors.

Mode $07 - Request On-Board Monitor Test Results

Mode $07 is used to request the status of on-board monitor test results from continuously monitored systems. This is also known in the industry as pending codes. Many states require that all of these monitors be passed in order for a vehicle to pass the state's emissions tests.

Mode $08 - Control Operations of On-Board Systems

Mode $08 allows the scan tool to control certain operations of the ECM. This is extremely handy when testing components and looking at system failures. Only advanced scan tools and some personal-computer-based scan tools have this capability.

Mode $09 - Vehicle Information

Mode $09 specifically queries the ECM for information about the vehicle. This information includes the VIN, the type of ECM in the vehicle, software calibration revisions loaded into the ECM, modules located on the vehicle, etc.

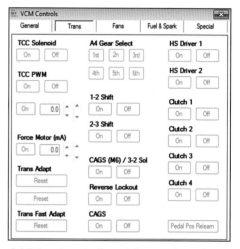

Additional control modules, such as the transmission, can also be controlled via the scan tool. (Screen courtesy HP Tuners)

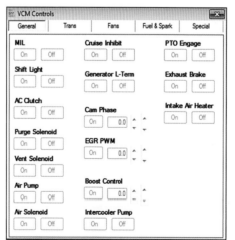

Most solenoids and emissions equipment can be controlled and cycled independent of conditions and commands. This is an excellent method for determining equipment failure. (Screen courtesy HP Tuners)

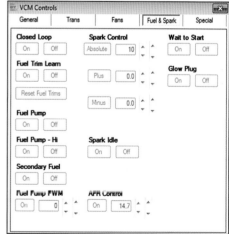

Real-time controls include the capability to reset learned trims, control air/fuel ratios, turn on and off closed-loop feedback, and manipulate cylinder spark timing. (Screen courtesy HP Tuners)

OBD-II MODES

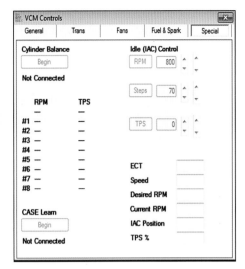

Some scan tools allow you to selectively kill cylinders, exercise the idle control hardware, and change engine speed. All of these tools are helpful in diagnosing an ill-running vehicle. (Screen courtesy HP Tuners)

Mode $09 shows the information about the ECM and a variety of control modules. In this example, a 2010 Chevrolet Camaro is scanned. The ECM (engine control module) and TCM (transmission control module) are shown. (Screen courtesy HP Tuners)

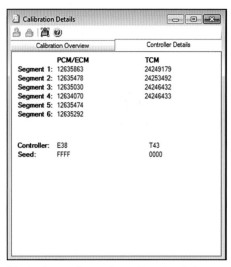

Mode $09 collects the calibration information stored with the control modules. This information can be compared against OEM TSB databases to determine if the software on the module requires an update. (Screen courtesy HP Tuners)

The Mysterious 10th Mode: $0A

For a long time there was a rumored 10th mode for OBD-II, which has since been confirmed and is now known as Mode $0A. This mode looks very similar to Mode $03 for confirmed codes and Mode $07 for pending DTCs. The key differentiator for Mode $0A is how stored codes are treated. If a drive cycle or scan tool clears a Mode $03 or Mode $07 code, the MIL is cleared. It is cleared along with traces of the codes themselves and the readiness tests being set to incomplete. Mode $0A now permanently stores these DTC codes until such a time that the readiness tests have been completed.

The main advantage of this new mode is that you can now see what readiness tests have not been completed. Then you can determine which DTC was set and cleared, and therefore prevented that readiness test from being completed. For example, say a DTC of P0442 was set for a faulty EVAP system module. A subsequent MIL was illuminated for the DTC. Suppose, in this scenario, the vehicle could not make it quickly enough to the technician before the DTC was cleared due to drive cycles. Without Mode $0A, the technician only saw that the readiness test was not completed. There was no evidence for what caused the test to not complete because the DTC cleared itself (or was cleared by a scan tool). With Mode $0A, you can see that a P0442 DTC caused the initial issue, and that information could help you pinpoint the faulty area in which to start the diagnostics.

CHAPTER 6

DIAGNOSTIC TROUBLE CODES

When OBD-I was first implemented, it offered the capability to alert the vehicle owner when the ECM detected a fault. When the fault occurred, an MIL illuminated somewhere on the dash to indicate that a problem had been found. Unfortunately, OBD-I was limited as to what it could detect as a fault. These faults were strictly limited to when a sensor failed and was no longer reporting values within a predefined, acceptable range. This means that no warnings were given about the degradation of systems until they actually had a catastrophic failure! This wasn't much of a preventative solution, which is what the EPA and CARB initially intended the system to be. Ideally, the system performed self-diagnosis on the vehicle and then alerted the owner that a serious failure was pending before it actually happened. Hence, the owner was compelled to service the vehicle and catastrophic failure was avoided.

OBD-II changed the capabilities of the DTC, along with the format standardization of the DTCs. Thanks to specification SAE J2012, the DTC was no longer limited to showing catastrophic failure when something finally quit working. With OBD-II, the ECM now not only checks for sensor failure, but it also constantly monitors the performance of sensors, as well as many vehicle systems, to ensure that they are operating within parameters of the EPA-mandated emissions laws. This includes efficiencies of the catalytic converters and other emissions-specific devices. It also includes tests, including engine misfires, VOC diurnal emissions from the fuel systems, and fueling system efficiencies of the engine. These checks alert the mechanic that, although sensors may not be at the failure point yet, there are certain parameters within the engine's operating specifications that are no longer being met. This allows an early interdiction before the problem worsens, negatively impacts fuel economy and emissions, and possibly increases the repair costs at a later time.

The latest additions to the checking capabilities of the OBD-II system incorporate checks on non-engine systems, including transmissions, safety systems, climate control systems, etc. There are predefined, generic diagnostic trouble codes that all manufacturers consistently implement. Each manufacturer also has defined specific DTCs for its vehicle lines. A quality scan tool has online diagnostic help, which interprets the DTCs and gives a briefly worded explanation of the code.

OBD-II Drive Cycle

The OBD-II system performs a diagnostic routine and monitors the results as the vehicle travels in a normal mode

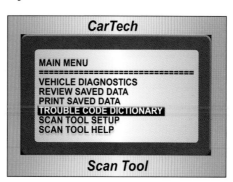

The scan tool may have a menu option, so that the mechanic can look up the various diagnostic trouble codes.

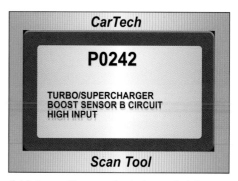

A high-quality scan tool gives a brief description of generic and manufacturer-specific DTCs.

of operation. The specifications describe a standard drive cycle that should exercise all of the systems so that they can be monitored. Normal day-to-day driving eventually satisfies the complete drive-cycle requirements. But occasionally, when attempting to reset a DTC, you may have to perform the specific steps to simulate a drive cycle. The generic drive cycle is described as:

1. The fuel tank is between 25- and 75-percent full.
2. A cold start where the engine coolant temperature is below 86 degrees F and allowed to warm up above 160 degrees F.
3. Accelerate to a speed between 40 and 55 mph and maintain that speed range for 5 minutes.
4. Decelerate without braking from 40 to 55 mph down to 20 mph, followed by a complete stop, having the engine idle for 10 seconds, and then turn the vehicle off and let sit for a minimum of 1 minute.
5. Restart the vehicle and accelerate to 40 to 55 mph with no greater than 25-percent acceleration and maintain this speed for 2 minutes.
6. Decelerate using brakes or clutch from 40 to 55 mph down to 20 mph, followed by a complete stop. Let the engine idle for 10 seconds, then turn the vehicle off, and let sit for a minimum of one minute.

Unfortunately, the generic description of the drive cycle is a limited description. Each manufacturer has tweaked the drive cycle to satisfy its requirements. For a complete drive cycle, along with the requirements, it is best to consult a service manual or search online for the details. For example, a General Motors drive cycle is defined as:

1. Perform a cold start to the engine where the engine coolant temperature is below 122 degrees F and the ambient air temperature is within 11 degrees F of the engine coolant temperature. The engine must be started immediately after the key is inserted into the ignition in order to not allow the oxygen sensor heaters to pre-heat the oxygen sensors.
2. The engine must idle for 150 seconds with the air conditioner on and the rear defroster on. If possible, turn the headlights on to help increase the electrical load on the vehicle. This allows the following monitors to pass: oxygen sensor heater, EVAP purge, air pump, and engine misfire. If the engine coolant temperature exceeds the closed-loop threshold, the fuel trim monitor will also be tested. The air conditioner should be turned off, and the vehicle accelerated to 55 mph with less than 50-percent throttle. Any monitors not tested from step number-two will be completed at this time.
3. The speed of 55 mph should be held steady for 180 seconds. During this phase, the following monitors are tested: oxygen sensor response, EGR, EVAP purge, engine misfire, and fuel trims.
4. Decelerate the vehicle by gradually coasting to 20 mph without using the brakes. At this time, the monitors for EGR, EVAP, and fuel trims are tested.
5. Accelerate using less than 75-percent throttle to 55 to 60 mph. This monitors oxygen sensor, EVAP, engine misfire, and fuel trims.
6. A steady speed of 55 mph should be held for 5 minutes. During this steady state, the monitors from previous steps attempt to complete, and therefore successfully finish their testing. In addition, the catalyst monitor diagnostics is performed. It may take up to five complete drive cycles for this monitor to be completed.
7. Repeat step number-five, again without using the brakes or clutch to slow down.

Why worry about a drive cycle test? Because certain monitors within the ECM require one drive cycle or more to be completed before the monitor is set. Many states require these monitors to pass testing, so the vehicle can complete the emissions test and a license obtained. (See more about monitors in Chapter 8.)

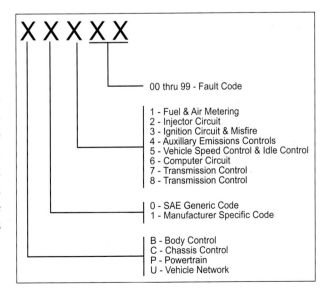

The standard SAE J2012 strictly defines the format of the DTC.

CHAPTER 6

Diagnosing and Repairing Body DTCs

Take care when working on any of the airbag systems because they are charged with an explosive and can be accidentally triggered during diagnosis and repair. The safety systems are best left to trained technicians at the dealership and professional repair shops. Furthermore, supplemental safety systems should never be rigged to circumvent or alter their intended operation.

Anatomy of a Diagnostic Trouble Code

The DTC, as defined by specification SAE J2012, has a very specific format that makes it very easy to dissect and understand. The DTC has five digits from four unique fields.

Field One

The first field defines the type of DTC: B, P, C, or U.

B: Body-generated diagnostic codes start with the letter B. These systems include safety restraint systems, such as side air bag controls, seat belt retention control, seat occupant sensors, etc.

P: Powertrain-generated diagnostic codes start with the letter P. These systems include engine components, cooling system equipment, transmission components, fuel system components, and emissions control equipment.

C: Chassis-generated diagnostic codes start with the letter C. These systems include active handling controls, traction control modules, anti-lock brake controller, and steering wheel sensors.

U: Network-generated diagnostic codes start with the letter U. These faults are generated when communication between modules is lost or interrupted. An example of this is a failure to communicate to the electronic dashboard control module.

Field Two

The second field determines if the DTC is a generic or manufacturer-specific code. If the second field contains the number 0, then it is understood that the DTC follows the SAE J2012 guidelines. If the second field contains the number 1, then the DTC is a manufacturer-specific DTC. Please note that manufacturers do not always adhere to a standard for their specific DTCs. Thus, a manufacturer-specific DTC for Honda may use the same manufacturer-specific DTC that Ford uses, but they may point to totally different fault types.

Field Three

The third field determines the system from which the DTC is originating. There are 10 possible systems that can generate the DTC, ranging from 0 to 9.

Field Four

The fourth field consists of two digits, ranging from 0 to 99. This field is the actual fault code that has been generated.

For instance, DTC P0301 tells you the following:

P = Powertrain fault
0 = SAE generic fault code
3 = Ignition system or misfire
01 = Cylinder #1 fault

"U" DTCs

The network-level diagnostic trouble code is sometimes difficult to diagnose. The complex networks that run throughout the vehicle and pass communication between the various modules are constantly monitored. If too many errors are determined, or if a module cannot be contacted, the ECM sets a network-related fault. Unfortunately, these codes can often be a result of scanning the vehicle with a scan tool device. The scan tool requests the attention of a certain module. At that time, the control module responds to the request of the scan tool. If, at the same time, another module sends a request that is ignored, it can set the network fault. At this point, the modules are basically now out of sync.

An easy fix for this exists. Many times, you can reboot the vehicle systems by disconnecting and reconnecting the battery. Attempt this only if a module is not responding properly or the vehicle is left in a limp mode.

DIAGNOSTIC TROUBLE CODES

Thus, the DTC P0301 indicates that there has been a misfire fault detected on cylinder number-1.

Appendix B lists an entire compilation of the generic SAE DTC codes, which can be used as a diagnostic reference.

Pending Diagnostic Trouble Codes

When the MIL is illuminated, a hard DTC has typically been set and stored in the ECM. Another type of DTC, known as a pending DTC, is also stored temporarily in the ECM when a fault is detected. If a fault is detected during a drive cycle, and this is not categorized as a severe fault, a pending DTC is set in the ECM. No freeze-frame data is saved for this fault. If the fault does not reoccur after one drive cycle, the pending DTC is cleared from the ECM. The MIL is not illuminated at any time when the pending DTC is set.

If the fault that caused the pending DTC to be set occurs again within the same or next drive cycle, then the pending DTC is cleared, and the hard DTC is set. The freeze-frame data for the fault is saved at the time of the occurrence.

Certain scan tools require a separate menu to read pending DTCs, and other scan tools mark the codes as "pending" or with a "P" beside them. Consult the scan tool documentation to determine which method your scan tool uses.

The pending DTCs are invaluable when trying to diagnose an intermittent problem. A lot of the test takes several failures to set a permanent DTC and illuminate the MIL. If a sensor is still functioning, but occasionally performing outside of its normal parameters and thus throwing performance or emissions off, it may not satisfy the permanent DTC requirements. Thus, the vehicle's owner may never know that there is a problem. And without the pending DTC, the mechanic may not be able to determine what sensor is intermittently causing the problem. With the pending DTC, the mechanic can determine that a sensor had an issue, but the issue did not continue. This information may point to a sensor or other hardware that is close to the end of its useful life, or may point to another issue intermittently affecting the sensor. A mechanic can never have too much information!

Diagnostic Trouble Code Types

There are four types of DTCs within the specification: Type A, Type B, Type C, and Type D. These types cover the array of component faults that allow for glitches or anomalies to appear in the testing, or to allow for non-emissions faults that do not affect performance or emissions standards. This gets back to the premise that the automobile is a dynamic platform that faces some of the harshest environments, and therefore an occasional glitch can be expected. (As a result, the DTCs allow for an occasional glitch for some sensors and hardware because glitches are expected.) Critical sensors for safety or emissions are dramatically affected even by the slightest abnormality and can be treated as a zero-tolerance test by the on-board diagnostics.

Type A

The most severe DTC type is Type A. An emissions fault generates this DTC type the first time that it is detected. Consider this the DTC for those critical-mission devices that cannot fail even once. When the first fault is detected within the device/sensor, the MIL illuminates. The freeze-frame information is stored to save the run-time data at exactly when the fault was detected. Examples of this type of DTC include "P0240 catalyst efficiency," "P0442 EVAP vent performance," and "P1336 CKP not learned." These DTC types rarely go away by themselves and usually require immediate repair.

Type B

The most common DTC type is Type B. An emissions fault also can generate this DTC type. When the fault is detected the first time in a single drive cycle, a pending DTC is set. If the fault is not detected during the next set amount of drive cycles, the pending DTC is cleared. If the fault is detected in the next set of drive cycles, the pending fault is cleared, but a permanent DTC is set in the ECM. This illuminates the MIL, and the freeze-frame data when the last fault occurred is stored in the ECM.

Examples of this type of DTC include "P0121 TPS sensor fault," "P0155 oxygen sensor fault," and "P0751 transmission shift solenoid fault." Sometimes these DTCs clear themselves after quite a few drive cycles, but normally, these should be diagnosed and repaired quickly.

Type C

A third type of DTC is Type C. This DTC is set after one drive cycle fault, but does not illuminate the MIL nor does it store the freeze-frame data. Unfortunately, many vehicles are on the road today with Type-C code DTCs set. Because no MIL is illuminated, there is no way to inform the driver of the code. Furthermore, many states pass a vehicle through emissions testing with Type-C codes in the ECM because they are non-emissions faults. Some vehicles that are equipped with diagnostic reports on the dashboard may not report Type-C DTCs. It is usually a good idea to quickly scan the vehicle each time the oil is changed, to see if any of these faults are currently stored and address them accordingly.

These are typically non-emissions-type faults that include "P0712 transmission temperature fault," "P0461 fuel-level sensor fault," and "P0531 air conditioner refrigerant fault."

Type D

The final type of DTC is Type D. This basically tells the ECM that the DTC is not checked at all. The reason for this type of DTC is for compatibility. Many times, the same software is used for a variety of vehicle platform combinations. Suppose a vehicle that normally comes equipped with an automatic transmission is purchased with the optional manual transmission. Instead of having separate software controlling the DTCs, the ECM's software merely sets the automatic transmission DTCs to type D, so that they are effectively ignored.

Current and Historical Diagnostic Trouble Codes

When a fault has been detected, a DTC has been set in the memory of the ECM and the MIL illuminates. At this point, this is considered a current DTC and on many scan tools is noted with a "C" behind the DTC code. Current DTCs also have their freeze-frame data stored in the ECM's memory. At this point, the scan tool gives you the option of erasing the DTC from the memory of the ECM. If it is erased, the DTC and its accompanying freeze-frame data are permanently erased.

After three ignition cycles, if the DTC no longer shows the fault and the DTC has not been erased from memory, the DTC moves from current status to historic status. On a scan tool, this DTC is noted with an "H" behind the DTC code. After the DTC has been moved to historic status and the MIL is extinguished, the freeze-frame data may be lost, depending on the manufacturer of the ECM. The historical DTC is kept in the memory of the ECM for 40 ignition cycles. If after these 40 cycles, the DTC has not re-faulted, the historical DTC is permanently erased from the ECM's memory.

The historical DTC is meant to help diagnose intermittent faults. When a DTC is set, access to a scan tool may not be available. Because intermittent issues can occur sporadically, using the historical DTCs stored in the ECM's memory is very helpful. Another benefit of the historical DTC is using them to see if they might be the cause of a current DTC's fault. This is the ECM's way of leaving a trail of bread crumbs showing a possible cascade of failures.

Plan Your Work and Work Your Plan

A common misconception is that the DTCs replace the intuition and understanding of the mechanic. If only it was as easy as reading the DTCs and replacing what they showed as generating a fault! The DTCs are still just another tool in the toolbox of the mechanic. A DTC that is showing a fault detected by a component may not be showing the component as the root cause of the issue, and the component may be functioning as it should.

In order to diagnose a problem, you need a good understanding of what DTCs really mean, what information sensors provide, and how a closed-loop feedback system works. I recommend mapping out a plan of attack based on the DTC's initial reporting, in order to provide a proper diagnosis and eventual repair. A simple yet effective rule is: Plan your work and work your plan.

Following is an example of a diagnosis on a vehicle that has the MIL illuminated. A quick scan of the system shows three codes stored—"P0300 misfire code," "P0171 fuel system lean bank 1," and "P0174 fuel system lean bank 2." Initially, it appears that the misfire code and the lean fuel system codes are unrelated, but our mechanical background reminds us that lack of fuel can cause a lean misfire. A quick check of the freeze-frame data for the misfire shows nothing out of the ordinary. Focus on the fuel delivery system to see if it is the culprit.

Lean codes show up as fuel trims set to a maximum positive value. Checking the freeze-frame data confirms that bank 1 and bank 2 show trims at 25 percent. The oxygen sensors are suspect because they are responsible for the feedback portion of the fuel system. Many mechanics may simply replace the oxygen sensors, and this could be a very expensive and unnecessary repair. I advise further diagnosis through empirical data generated from the OBD-II system.

Using the real-time data recording, you can observe the voltage values of the oxygen sensors (and you need to understand what a failed oxygen sensor reports). But in this case the voltage values indicate that the oxygen sensors are performing within their parameters. Thus, the DTC indicating a lean fuel

Erasing DTCs and Monitors

When erasing DTCs, remember that system monitor tests are reset. Many states rely on the system monitors to have passed in order for a vehicle to pass their emissions tests. Because many of these system monitors require multiple drive cycles to pass, do not clear DTCs and immediately take the vehicle to the emissions testing station; some of the monitors will not yet be reset.

system is really pointing to another problem that is hidden at the moment.

The next step is to walk through the entire feedback circuit for the vehicle's fuel system and, on a piece of paper, verify that the components that report back to the ECM are within spec and in agreement with one another. But after looking at the various sensors and comparing them to assigned values, you note that the mass airflow meter (MAF) is not reporting the correct values, although it is still passing its own diagnostics at startup. The MAF dictates the airflow coming into the engine, and therefore sets up the amount of fuel the injectors will flow. If the MAF airflow is reported out of spec, the fuel will bias incorrectly. As the rest of the feedback system tries in vain to keep the system in balance, it fails and thus components that are operating normally start to set DTC faults

It all comes down to the old axiom: garbage in, garbage out. If a sensor is giving the system incorrect information and its problems are not detected, downstream sensors balance the system based off of the information given and fail, even though there is nothing wrong with the sensor in question. In this example, the MAF was dirty from having an oil-based air filter installed on the car, which coated the sensor and caused it to report airflow incorrectly. The fix was a simple cleaning. Unfortunately, mechanics often replace perfectly good sensors. In this case, a pair of oxygen sensors could cost hundreds of dollars and still not fix the issue.

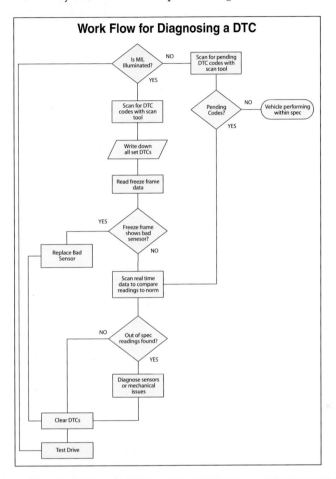

When faced with a DTC or pending DTC, it is important to logically attack the problem. This way, you can quickly and effectively identify the problem as well as avoid wasting hours and money on a problem that may not require a resolution.

Interpreting Diagnostic Trouble Codes

Most scan tools and trouble code readers have a built-in help mode, so you can decipher the DTCs stored in the ECM. There are some scan tools that don't have the online help or, more often, someone is describing a DTC and the scan tool is not handy. Fear not though, there are myriad sources to find the DTCs. Some options are:

- Consult the appendix of this book for generic DTC definitions.
- www.trouble-codes.com has an excellent list of generic and manufacturer-specific DTCs.
- www.obd-codes.com not only has codes but also diagnostic information and specific reasons why the code was set.
- The local public library may have diagnostic software by Mitchell or All Data, which helps to diagnose the code and give steps to verify the fault.
- NASTF. The National Automotive Service Task Force is a cooperative effort among the automotive service industry, the equipment and tool industry, and automotive manufacturers to ensure that automotive service professionals have the information, training, and tools needed to properly diagnose and repair today's high-tech vehicles.

CHAPTER 7

FREEZE-FRAME DATA

When the DTC is triggered, Mode $02 specified by SAE J1978 records it. In addition, it stores the sensor data and the calculated data at the time when it was set. At a minimum, the following data points are stored:

- DTC that was triggered
- Engine speed registered in RPM
- Engine load that has been calculated by the ECM
- Short-term fuel trim(s)
- Long-term fuel trim(s)
- Intake manifold pressure reading
- Vehicle speed in kilometers per hour
- Coolant temperature
- Fuel system status

As discussed earlier, Mode $02 is the same data structure as Mode $01 ("show current data"), except for PID 02. PID 02 is returned in the data stream and adds the trouble code to the sensor/monitored data. Unfortunately, there are data sets that often appear in Mode $01 that are not available in Mode $02, so not all of the sensor/calculated data may be available in the freeze-frame data. Fortunately, the pertinent data helps to lead you to the proper diagnosis. As technology continues to improve, more data is stored in the freeze-frame data set available to you. Some vehicles even store two snapshots before the error occurred: A snapshot of the data right after the error, and then two more snapshots a few seconds after the error occurred.

Freeze-Frame Data Reports

Freeze-frame data is the primary path to diagnose a problem on the vehicle. At a minimum, the ECM captures a snapshot of the running data right after the DTC was detected and set. The data set is tied to the specific DTC, so if there are multiple DTCs currently set or pending, the ECM has a data snapshot of the conditions when each DTC was established. The freeze-frame data consists of as many as seven different reports:

- Freeze-frame data summary
- ECM information
- System readiness test results
- Specifically monitored system test results
- Oxygen sensor test results
- DTCs currently established or pending
- Historical data

Test results for the system readiness tests, specifically monitored tests, and the oxygen sensor tests are merely reports of the working status of these tests. The oxygen sensor tests have been outdated for several years; so many newer vehicles do not support this test at all. The system readiness tests show the current status of the tests for the systems that are critical to emissions. Typically, if a DTC is established and the MIL is illuminated, the corresponding system readiness tests will not be passed.

Freeze-Frame Data Summary

The freeze-frame data summary is a brief report on the status of the communications to the ECM. Figure 7.1 contains the following information:

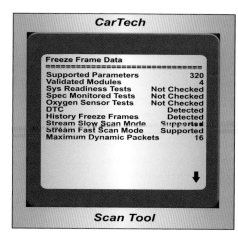

Figure 7.1. This scan-tool screen summarizes the status of the freeze-frame history logs.

FREEZE-FRAME DATA

Supported parameters available: 320

The freeze-frame data displays the total number of PIDs that are available. Keep in mind that not all of the PIDs may contain freeze-frame information.

Validated modules available: 4

This particular vehicle contains four control modules, which were scanned. If the error is generated from outside the ECM, the flagging module is shown in the error message.

System readiness test (EPA): Not checked yet

This vehicle has not had its readiness tests polled or checked at this time.

Specifically monitored tests: Not checked yet

Specifically monitored tests on this vehicle have not been polled or checked at this time.

O_2 sensors tests: Not checked yet

If supported, the oxygen sensor tests have not been polled or completed. It is important to remember that if a vehicle does not support a particular test, especially the outdated oxygen sensor test, it will not show that check as completed but still may show the test available in the scan tool.

Diagnostic trouble codes: Detected

One or more DTCs have been established and saved in the scan tool data made available from the ECM.

History freeze frames: Detected

One or more of the DTCs that have been established contain historical freeze-frame data.

Historical Freeze-Frame Data

The historical freeze-frame data is stored in the ECM's memory of running conditions immediately after the DTC was established. Certain vehicles can store more than one historical freeze-frame data set, which can greatly aid in diagnostics for the vehicle.

Depending on the scan-tool manufacturer, you may be able to print out the entire freeze-frame data on a connected printer, or download it to a personal computer to be printed. In the

Units of Measurement

The automotive industry spans the world but, unfortunately, there is no world-wide, uniform measurement system in use. In the United States, for example, distance is typically measured in miles, while most areas of the world measure distance in kilometers. Many countries' scientific and engineering communities have adapted to the International System of Units (abbreviated as SI) measurement system, yet most countries have not eliminated all other forms of measurement (i.e., inches, pounds, pounds per square inch, etc). This can create some confusion when looking at the units reported from a scan-tool device.

The SI is the modern form of the metric system and was developed in 1960. It uses seven base units and powers of 10, based on their prefixes to the base unit. For example, the prefixes used by a scan tool include:

- Gigahertz   1,000,000,000 Hertz
- Megabit   1,000,000 bits
- Kilometer   1,000 meters
- Centigram   0.01 gram (100 centigrams in 1 gram)
- Milliliter   0.001 liter (1,000 milliliters in 1 liter)
- Nanosecond   0.000000001 second (1 billion nanoseconds in 1 second)

Unfortunately, not all ECMs conform to one standard. Luckily, most higher-end scan tools convert units of measurement to the current application's standard.

Common measurement units include:

Measurement	SI Units	Multiply By	U.S. Equivalent
Length	m	3.28084	ft
Distance	km	0.621371	miles
Area	m^2	10.7639	ft^2
Volume	m^3	35.3147	ft^3
Velocity	km/h	0.625137	mph
Pressure	kPa	0.145037	psi
Liquid Volume	liters	0.26417	gallons
Temperature	degrees K	$9/_5 - 459.67$	degrees F
Temperature	degrees C	$9/_5 + 32$	degrees F
Weight	grams	0.002204	pounds
Torque	Newton	0.73756	ft-lbs

CHAPTER 7

OBD-II System Measurements

When the scan tool queries the ECM, the ECM reports all measurements, but it is always important to understand the units of the measurement. As informative as the data can be during a diagnostic procedure, you must always be careful to note the units that are being reported as to not confuse the data values or possibly misdiagnose a problem due to a misread value. The scan tool always reports the units of measurement, so a general habit should be to always look at the units and, if necessary, change or convert them to units that are most applicable.

Freeze-Frame Data	Value	Explanation	Status
Fuel System Bank 1	CL	Denotes that the fuel system for bank 1 is in closed loop	Desired
Fuel System Bank 2	CL	Denotes that the fuel system for bank 2 is in closed loop	Desired
Calculated Load Value	3.1 percent	The engine load was at 3.1 percent (idle loads) when the DTC set	Desired
Engine Coolant Temperature	107° C	The engine was at 107°C (225 degrees F), which is at operating temp	OK
Short-Term Fuel Trim Bank 1	2.3 percent	Instantaneous fuel correction on bank 1 is adding 2.3 percent of fuel	OK
Long-Term Fuel Trim Bank 1	0.8 percent	Historical fuel correction on bank 2 is adding 0.8 percent of fuel	OK
Short-Term Fuel Trim Bank 2	14.8 percent	Instantaneous fuel correction on bank 2 is adding 14.8 percent of fuel	Concern
Long-Term Fuel Trim Bank 2	25 percent	Historical fuel correction on bank 2 is adding 25 percent of fuel (and caused the DTC)	Concern
Intake Manifold Absolute Pressure	59.0 kPa	Given that the load was 3.1 percent, this reading is a bit high	Concern
Engine RPM	0 kmh	Vehicle speed indicates that the vehicle is stopped	OK
Airflow rate from MAF	9.11 g/s	Air flow reading from the MAF is 9.11 grams per second	OK
Absolute Throttle Position	0.0 percent	The throttle is showing as closed which correlates to the engine and vehicle speed	OK
TCC PWM Solenoid	Off	The transmission is not in lockup mode	OK
EGR Valve Sensor Voltage	0.0v	The EGR valve is currently closed and not affecting the air/fuel ratio readings	OK
ECT at startup	121° C	The vehicle was started up in a warm condition	OK
Desired Idle Speed	700	The ECM is trying to command 700 rpm at idle	OK
Barometric Pressure	97.8	Typical pressure is around 100 kPa	OK
Commanded A/F Ratio	14.6:1	Stoichiometric ratio on a normal-running vehicle	OK
Engine Run Time	1039	The DTC was thrown after about 17 minutes	OK
Misfire Count Status	255	There have been 255 logged misfires in the PCM This is a fairly low number and of no concern	OK
Mileage since last code cleared	145 km	It has been a short time since the last DTC was cleared, which indicates a continuing problem	Concern
Odometer when last code cleared	200 km	It has been a short time since the last DTC was cleared, which indicates a continuing problem	Concern
Injector Base Pulse Width Bank 1	2.045 ms	Bank 1 injector pulse width indicating how much fuel is added	OK
Injector Base Pulse Width Bank 2	2.609 ms	Bank 2 is significantly longer than bank 1 indicating a possible issue	OK

FREEZE-FRAME DATA

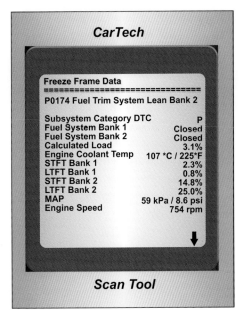

Figure 7.2. This is a partial listing of the data reported under the freeze-frame data menu. Typically, this information traverses several screens, so it is important to print out the freeze-frame data if the scan tool supports it, or use the freeze-frame datasheet included in the appendix of this manual.

event that the scan tool does not support these capabilities, a freeze-frame diagnostic paper is included, which can be photocopied and used during diagnostics.

Figure 7.2 is an example of historical freeze-frame data that was saved in the ECM's memory when a P0174 DTC was established. Explanations of the readings and what you might look for in this particular instance are included. This is meant as a starting point for looking at a variety of issues, which may lead to the DTC being set.

Breaking Down Freeze-Frame Data

The freeze-frame data gives a snapshot of the conditions immediately after the DTC was established, although some vehicles may give multiple before-and-after snapshots. There are three key areas within the multitude of information that the freeze-frame data provides:

- Sensor feedback
- Operating conditions
- Calculated values

You should concentrate on examining the sensor data. It is imperative to understand the sensors that are equipped on the vehicle, their functions, and their valid data values. As you page through the freeze-frame data, look for any sensor values that are outside of their normal reporting range, such as an intake air temperature sensor reporting a temperature of -48 degrees F on a warm summer day. Also, look for conflicting data from different sensors, such as a throttle position sensor reporting a 100-percent throttle level though the mass air flow meter only shows idle levels of airflow. Finally, look for sensors that may be reporting invalid data, such as an oxygen sensor that is only swinging to a high point of 550 mV and a low of 400 mV. (Chapter 15 covers specifics on a variety of common sensors and helps to explain the data that the sensors provide.)

Diagnostic Information

"Information is not knowledge."
– Albert Einstein

This quote encapsulates automotive diagnostics. In essence, the scan tool can provide most of the information that is required in the diagnostic of an automobile, but it is up to the technician to be able to dissect, discern, and react to that information. It is up to the knowledge of the technician to know where to look and what to do with the scan tool data in order to understand the problem.

DTC P0121

Throttle/Pedal Position Sensor/Switch A/Circuit Range/ Performance Problem

Detection Condition. If the PCM detects that the throttle valve opening angle is below 12.5 percent for 5 seconds after the following conditions are met, the PCM determines that the TPS is stuck closed:

- Monitoring condition
- Engine coolant temperature is above 68 degrees C
- MAF sensor signal is above 30 grams per second

Detection Condition. If the PCM detects that the throttle valve opening angle is above 50 percent for 5 seconds after the following conditions are met, the PCM determines that the TPS is stuck open:

- Monitoring condition
- Engine speed is above 500 rpm
- MAF sensor signal is below 5 grams per second

CHAPTER 7

Operating conditions and calculated values are other areas of the freeze-frame data that are studied. You often need to understand exactly what was going on from a performance standpoint when the error occurred. Was the vehicle idling, driving under heavy load, or at wide-open-throttle (WOT) operation? How long was the vehicle running before the code was set? These data points combined with the sensor data can help isolate the specific conditions that may be causing the problem.

The sidebar "Diagnostic Information" on page 51 shows a portion of the diagnostic aid for a particular DTC, in this case, P0121. This standard diagnostic aid chart describes a lot of the information shown in the freeze-frame data and is critical to reference when diagnosing an issue. First and foremost, diagnostic information shows the operating conditions that must be obtained so the DTC can be detected. A quick check of the freeze-frame data can confirm if this condition was obtained, and it is also a good time to ensure that the sensors that are responsible for establishing this condition are reporting valid results.

The next piece of key information to be compared between the diagnostic aid and the freeze-frame data is contained in the Monitoring Conditions sections. For this particular DTC, the conditions that are monitored include information from the engine coolant temperature sensor, the mass airflow meter, engine speed in RPM, and finally, the information coming from the TPS (throttle position sensor). By analyzing the data reported in the freeze-frame information and comparing the thresholds for the data in the diagnostic aid, you can identify if the throttle position sensor is the issue, or possibly one of the sensors that is used to monitor the TPS is the culprit.

Using the Freeze-Frame Data Example

Using the historical freeze-frame data in the previous example, there are a few areas of concern that can prove the established DTC P0174, as well as lead to other areas that should be investigated. Five pieces of information in the freeze-frame data are labeled "CONCERN," and an experienced technician investigates those problems.

First and foremost, because the DTC indicates that there is a lean condition on bank 2, the short-term fuel trims and long-term fuel trims for that bank should be studied. The freeze-frame data shows that this bank is very lean (25 percent) and thus set the DTC. It is also noted that bank 1 fuel trims are within specifications, so that eliminates a problem with the fuel supply, which would appear on both banks. Next, it is noted that the manifold pressure reading is abnormally high. Typically, this reading at idle (as indicated in the freeze-frame data) should be near 30-kPa. Because the reading is almost double at idle, this indicates that either the sensor is not returning valid data, or there is a vacuum leak somewhere in the engine. Finally, the mileage shown between cleared DTCs indicates that other DTCs have been set in the past and have been cleared. Hence, there may be preexisting, or multiple occurrences of the DTC.

In this example, further investigations found a slight intake leak on bank 2, which was causing excessive unmetered air to enter the engine, and this showed up as a lean condition via the oxygen sensors. Since the MAF did not detect the air, the ECM did not account for this extra air. Replacing the intake gasket on that side of the engine cured the problem. Without the information from the manifold pressure sensor, many technicians blame a bad oxygen sensor and replace that part unnecessarily. This wastes time and money on a part that may have been functioning correctly, and it would not fix the physical problem that was present on the vehicle.

Banks and Sensor Counts

Scan tools and auto manufacturers refer to the vehicle in halves, which are labeled as banks. The banks are labeled as bank 1, bank 2, bank 3, and so on. Odd-numbered banks are on the driver's side of the vehicle and even-numbered banks are located on the passenger's side of the vehicle. So a DTC "P0171 System Lean Bank 1" denotes that the driver's-side bank shows that its air/fuel ratio is too lean.

On some systems, multiple sensors may be present on a single bank. The order of the banks determines its position with respect to the front of the vehicle, with number-1 and number-2 sensors located closest to the front. For example, "oxygen sensor bank 1 sensor 2" (often abbreviated as B1S2) denotes the oxygen sensor located on the driver's-side bank, and is second in line.

You can trace the exhaust system, starting at the cylinder head, and locate the oxygen sensors and count them as the exhaust moves away from the cylinder head.

CHAPTER 8

EMISSIONS TESTS AND MONITORS

Because the OBD-II system is designed to verify the emissions equipment on an engine and vehicle, many states do not conduct tailpipe emissions tests and instead carry out a pure plug-in test. The plug-in tests rely on reports from the ECM on the status of the emissions equipment on the vehicle. The system monitors are basic tests of emissions-control components that the ECM performs constantly or under certain circumstances, based on the component being tested. The ECM retains the results, or lack thereof, for all the monitors and reports the test results directly to the scan tool.

Emissions Tests

When a vehicle is brought to a state-certified emissions test, the first thing that the system checks for is an illuminated MIL. Regardless of the reason for the DTC, the vehicle cannot pass the emissions test with one set. The system also tests to see if the light works; so, all of you out there with your nippers, you can't just cut the light out of the dashboard. The emissions tester typically cycles the ignition key to the run position, to ensure that the MIL is illuminated when the engine is not running and that the light extinguishes once the engine is started.

If the emissions diagnostic computer cannot communicate with the ECM, the vehicle also fails the test. Most of the time, the fuse that controls the DLC port (and also the cigarette lighter) is blown. Unfortunately, most emissions technicians automatically fail the car and require the owner to remedy the problem. But if you have a scan tool, all of the tests that the emissions technicians run can be verified before actual testing, to ensure that the vehicle passes the test. If your scan tool won't connect, grab your VOM and check for voltage on the DLC port. If there is no voltage present, check the fuses marked "ACCY" or "CIG" in the fuse block.

In order to pass most state emissions tests, the following criteria must be met:

- The MIL must illuminate when commanded.
- During the engine run test, the MIL must not be illuminated.
- All system monitors are ready and have passed.
- There are no DTCs set in the computer, although expired DTCs are accepted.

System Monitors

The ECM is required to run a bank of tests on the components used in the engine controls. This is to verify that they are working properly in order to minimize the emissions emanating from the vehicle. These tests are known as system monitors or monitors for short. In order to not bog down the ECM with a lot of random tests, there are nine major monitors and three minor monitors.

The major monitors (required for emissions testing if applicable to vehicle) are:

- Engine misfire monitoring
- Evaporative system monitoring
- Heated catalyst monitoring
- Secondary air system monitoring
- Fuel system monitoring
- Oxygen sensor monitoring
- EGR system monitoring
- Comprehensive component monitoring (CCM)
- Catalyst efficiency monitoring

The minor monitors (not required for emissions testing) are:

- Air conditioning system monitoring
- Positive crankcase ventilation monitoring
- Thermostat monitoring

The ECM groups a set of sensors into a specific monitor to test a particular

CHAPTER 8

subsystem. Please note that each monitor is run during a specific set of operating conditions and, therefore, every monitor may not be run during a quick normal drive cycle. In fact, some of the monitors take a very long time to test due to their required operating parameters. It is for this reason that many states allow one or two monitors to be incomplete during testing and still pass the vehicle.

Monitors are divided into two different classes: those that run constantly and those that run when specific operating conditions are met. The monitors that run constantly are known as "continuous," and those that are run only when specific conditions are met are known as "non-continuous." Obviously, the continuous monitors are much easier to get set to "ready" status because they are constantly being monitored. The non-continuous monitors are trickier because themechanic must know the specific drive cycle to perform in order for the monitor's conditions to be satisfied.

Resetting Monitors

It is very critical to remember that when the ECM is reset or a DTC is cleared, all of the monitors are reset and their states are set to "not ready." So if a mechanic clears a DTC and then attempts to test the monitors, the vehicle will fail the emissions test because most of the monitors will not be set to "ready" yet. In most cases, very particular drive cycles must be performed in order to get the monitors to set themselves to "ready" in order for the vehicle to pass emissions. This is why it is critical to not wait until the last minute to test the vehicle, and why the DTCs should be cleared and diagnosed well ahead of the emissions inspection. Simply reading DTCs does not affect the status of the monitors.

Keep in mind, the most important OBD-II pieces of equipment on the vehicle are the catalysts, so the ECM is always continuously testing these monitors: engine misfires, fuel system, and comprehensive component. Any system or component failures that are monitored in these tests may harm the catalyst, thus they are always being run. The rest of the monitors are only tested when the criteria for the test is met because failure within their system or sensors do not place the catalyst in jeopardy.

Monitor Modes of Operation

The ECM assigns one of four modes of operation to each monitor. These operational modes are:

- Normal
- Pending
- Conflict
- Suspended

When things are working properly, all of the monitors are in a normal mode. If problems arise, the ECM may place one or more monitors into one of the three other modes. When a sensor or system has malfunctioned, the ECM places that particular monitor into a pending mode. While in this mode, the ECM will not set the monitor to Ready until the problem has been resolved or components have been repaired.

The ECM has a particular order in which the monitors must be run. Thus, if one monitor is not completed, the next monitor may not be run until the first one is completed. If this is the case, the monitors that are waiting for the first one to complete are placed into a suspended mode. Once the problem monitor is completed, the rest of the following monitors move to normal mode. If there is a case in which two different monitors are trying to perform the test at the same time, then one of the monitors is placed into conflict mode while the other monitor finishes its testing. When that testing is complete, then the conflicted monitor is placed back to normal and allowed to finish. Remember that all of this testing is going on simultaneously, while the ECM is controlling the entire drivetrain going down the road!

Misfire Monitor

The misfire monitor is probably the most critical monitor when it comes to protecting the catalyst because misfires can overload the catalyst with unburned fuel, which not only reduces its efficiency but can physically damage and render the catalyst ineffective. As a best-case scenario, a slight misfire increases hydrocarbon emissions. This is why this particular monitor is a continuous monitor.

Monitor Testing Criteria
- Engine coolant temperature must be within a specific range.
- Engine speed must be within a specific range.
- Vehicle speed must be within a specific range.
- Valid data from manifold pressure sensor and airflow calculations or mass air flow meter readings.

Monitor Modes of Operation
- Normal should be achieved if the testing criteria is met.
- Pending is assigned if any of the sensors produce invalid data or have a DTC assigned to them, including vehicle speed, engine RPM, ECT, MAP, and airflow through calculation or MAF.
- Conflict is assigned if the fuel trims are outside their normal lean/rich range, the evaporative system has failed, or the exhaust gas recirculation system has failed. Any of these system failures causes a fueling

problem, which can also cause a misfire issue.
- Suspending is assigned if the engine is still being started or has just started in a cold start mode. It can also be suspended if the data from any of the sensors is showing occasional improper readings and if the fuel level is too low, which may contribute to inadequate fuel pressure and/or a lean fuel delivery condition.

Evaporative System Monitor

Before OBD-II was installed on vehicles, fuel-vapor emissions monitoring was essentially non-existent. Volatile organic compounds (VOC) escape from the fuel in the fuel system. VOCs enter the atmosphere through the evaporation of fuel. According to the United States EPA's *Automotive Emissions: An Overview*, published in 1994, a majority of the VOCs present in an air sample today are a direct result of the evaporation of fuel from vehicles.

The four main sources of VOC emissions are:

- Refueling emissions
- Diurnal emissions
- Running emissions
- Heat soak emissions

In many places across the United States, mandates are in place to capture refueling emissions at the fuel nozzle. This source of emissions is outside the realm of the OBD-II system. The other sources of emissions are essentially within the design of the vehicle and the fueling systems. So, what part does OBD-II play in limiting VOC emissions?

The fueling system on modern vehicles is a closed, pressurized system. In other words, it no longer has vents tied directly to atmosphere, which allows VOCs to escape. Instead, a system of valves, solenoids, and sensors tracks and releases accumulated vapors into the engine, where they are burned in the combustion cycle. The OBD-II system uses sensors to monitor the presence of pressure or vacuum and can then actuate the valves and solenoids to ensure that the system is not leaking.

A simple example of this is when a person forgets to put the fuel fill cap back on or leaves it loose. The OBD-II system sees that it cannot hold pressure in the system, and it illuminates the MIL on the dash indicating an issue. Furthermore, when the valves are actuated and vapor is introduced into the engine's combustion cycle, the ECM should see an increase in activity from the exhaust air/fuel sensors. Failure to see this activity is another way that the ECM can determine that there is a fault in the evaporative system.

Monitor Testing Criteria
- Engine coolant temperature must be above closed-loop enable temperature (about 160 to 170 degrees F).
- Oxygen sensors must be at temperature and functional.
- Engine is running in closed-loop feedback mode.
- Throttle position and engine speed are within a specified range.
- Transmission is in a specified gear or gear range.

Monitor Modes of Operation
- Normal should be achieved if the testing criteria is met.
- Pending is assigned if any of the sensors produce invalid data or have a DTC assigned to them, including oxygen sensors, engine speed pickup, engine coolant temperature, throttle position sensor, or if the transmission has been placed in a gear outside the specified range.
- Conflict is assigned if the oxygen sensor monitors are running or fuel system monitors are running.
- Suspending is assigned if the oxygen sensor monitor has failed or the fuel system monitor has failed.

Heated Catalyst Monitor and Catalyst Efficiency Monitor

The single most important piece of emissions equipment in terms of clean air quality is the catalytic converter. The catalyst contains a honeycomb brick of materials that includes precious metals and ceramics. This brick is heated and performs a chemical reaction on the incoming exhaust stream. It converts the carbon monoxide, hydrocarbons, and nitrogen oxides in the exhaust emissions into water vapor and other less-harmful gasses. As long as the health of the catalyst is intact, this process works flawlessly.

OBD-II's job is to ensure that emissions are kept to a minimum by checking that the catalytic converter is performing its duty. To do that, two oxygen sensors are employed per catalytic converter: One is placed before the catalyst (pre-cat) and the other is placed after the catalyst (post-cat). The pre-cat sensor is used to drive the fuel-system correction, while the post-cat sensor is used to monitor the operating efficiency of the catalyst. The premise is that the pre-cat sensor shows the common sinusoidal fluctuations from rich to lean and back again, while the post-cat sensor shows a consistently low level of hydrocarbons because the catalyst has scrubbed much of them from the exhaust stream. If the post-cat sensor is picking up fluctuations, then the ECM can assume that the catalysts are no longer functioning properly and a DTC may be set.

Monitor Testing Criteria
- Engine coolant temperature must be

above closed-loop enable temperature (around 160 to 170 degrees F).
- Vehicle is not idling based on throttle position readings.
- Engine is running in closed-loop feedback mode.
- Engine speed and airflow readings are within a specified range.

Monitor Modes of Operation
- Normal should be achieved if the testing criteria is met.
- Pending is assigned if any DTCs related to engine misfire, fuel system, or oxygen sensor have occurred or if the vehicle cannot maintain closed-loop operations.
- Conflict is assigned if the engine has not gone through its warm-up cycle or if the EGR, fuel system, or evaporative emission control (EVAP) monitors are in operation.
- Suspending is assigned if the oxygen sensor monitor has failed or the fuel system monitor has failed.

Secondary Air System Monitor

The original materials that catalysts contained took a bit of time to reach their operating temperature, at which the catalytic conversion could occur. In order to speed this heating process, vehicle designers implemented a methodology that injected air into the exhaust stream to facilitate a quicker, hotter burn of excess fuel. This helps to bring the catalyst up to temperature. A simple air pump is installed before the catalyst, and the ECM actuates it while warming up the catalyst.

The ECM monitors the oxygen sensors for activity. When they are warmed and reporting that the exhaust air/fuel ratio is lean from the air pump injecting air, the ECM can shut off the air pump. After the air pump is shut off, the closed-loop operation of the ECM should show the oxygen sensors toggling from rich to lean and vice versa as they are supposed to. If this toggling is present, the secondary air system monitor is then passed.

The ECM may also test the air pump by turning it on at a random time and monitoring the results from the oxygen sensors. If the sensors do not read a lean condition when the air pump is on, then the ECM indicates that there is an issue with the air pump or valves that control it.

Since the catalyst materials have changed in recent years, not all vehicles are equipped with a secondary air system. On these vehicles, this test is not run nor is it valid.

Monitor Testing Criteria
- ECM starts in open-loop operation.
- ECM must switch to closed-loop operation.
- Pre-cat oxygen sensors returning valid readings.

Monitor Modes of Operation
- Normal should be achieved if the testing criteria is met.
- Pending is assigned if the engine cannot enter closed-loop operations or the comprehensive component monitor has a set or pending DTC.
- Conflict is assigned if the heated catalyst monitor is running or the oxygen sensor monitor is running.
- Suspending is assigned if the oxygen sensor monitor or oxygen sensor heater circuit has failed, or the engine is shut off before the ECM can enter closed-loop operation.

Fuel System Monitor

Like the misfire monitor, the fuel system monitor is a continuous monitor, which can have detrimental impact on the catalytic converters. Unlike the misfire monitor, the fuel system monitor has to take into account many operating variables of the engine in order to ensure that one particular function is not skewing the results. The monitor's main function is to ensure that the adaptable fuel system responds properly to the demands from the feedback of the oxygen sensors. Mainly, the monitor is watching the fuel system trims to make sure that they are responding in a proper way.

Monitor Testing Criteria
- Engine coolant temperature must be above closed-loop enable temperature (around 160 to 170 degrees F).
- ECM is in closed-loop operation.
- Valid data gathered for engine coolant temperature, engine speed, intake air temperature, manifold pressure, throttle position, and vehicle speed.
- Long-term and short-term fuel trim values are valid and present.

Monitor Modes of Operation
- Normal should be achieved if the testing criteria is met.
- Pending is assigned if a DTC is set for the oxygen sensor, the EVAP monitor has failed, or the EGR monitor has failed (these all put the air/fuel ratio in question).
- Conflict is assigned if there are any pending DTCs from engine misfires, EVAP system, oxygen sensors, or EGR monitor.
- Suspending is assigned if the level of fuel in the fuel tank is insufficient or any of the sensor data is determined to be outside the scope of the test.

Oxygen Sensor Monitor and Heated Oxygen Sensor Monitor

The oxygen sensors are the main feedback device for the ECM when it comes to evaluating the air/fuel ratio of the exhaust stream post-combustion

EMISSIONS TESTS AND MONITORS

cycle. Without being able to gain the information of this air/fuel ratio, the ECM cannot adjust the fuel delivery, and may end up creating more emissions, and possibly harming emissions and non-emissions equipment. Therefore, these tests are continuously monitored to ensure the health and well being of the entire emissions system.

When checking the oxygen sensors, the ECM pays particular attention to the voltage levels that the oxygen sensors return. What the ECM expects to see is smooth, sinusoidal-shaped data that switches from rich to lean fairly consistently and has a switch point of approximately 450 mV. The ECM checks the frequency of the switching signal to make sure that the response time from the sensor is within spec.

Because the oxygen sensor relies on heat to start the chemical reaction that reports back the air/fuel ratio, it is important that the sensor comes up to temperature quickly, and maintains its heat in a suitable range. The ECM can check the operation of the heaters inside the oxygen sensor as it cools down after the engine is shut off.

Monitor Testing Criteria
- Engine coolant temperature must be above 120 degrees F.
- Engine has to be running for a predefined amount of time.
- Throttle position must be within a specified range.
- EVAP system must not be in a purge cycle.
- Transmission must be in specified gear range.
- Vehicle speed must be within specified range.

Monitor Modes of Operation
- Normal should be achieved if the testing criteria is met.
- Pending is assigned if any DTCs are set for engine misfires, oxygen sensor-related faults, and vehicle speed sensor faults.
- Conflict is assigned if the fuel system monitors are running, pending DTCs are set for misfires or oxygen sensor-related faults, or the engine has not had enough run time yet.
- Suspending is never assigned because there are no other related tests that can take precedence over the oxygen sensor monitors.

EGR System Monitor

The EGR system is integral to emissions and combustion temperatures because it returns some of the exhaust gases back into the combustion process to be re-burned. Not only does the EGR system bring down the NO_x emissions, it also helps to cool the combustion chamber, which aids in preventing detonation or spark knock.

Unfortunately, the art of getting the proper amount of EGR actuation is a bit tricky. The ECM can control the timing of the valve that opens to allow the exhaust gases back into the intake track, and it can control the exact operating conditions for when these events can occur. But the ECM cannot track the effectiveness of the EGR system on emissions quality through a specific sensor reading that's dedicated to do so. However, manufacturers can use several tricks or operations. They can see if a particular sensor can detect a difference in manifold pressure or a change in exhaust air/fuel ratio when the valve is open, or add in a dedicated sensor to detect an increase in heat or pressure. Thus, even though the ECM cannot directly tell if the EGR is impacting emissions, it can at least tell if it is functioning.

To check the functionality of the EGR system, the ECM commands the valve to cycle while watching a sensor that the manufacturer has assigned to detect the change in flow. The ECM chooses a time when the EGR is not active and then tests it for this monitor. This test happens all the time while the vehicle is running.

Many newer vehicles are no longer equipped with an EGR system. Advances in engine design and camshaft manipulation allows the engine designers to eliminate the EGR requirement. Vehicles not equipped with an EGR system do not run this monitor. From an emissions testing standpoint, this monitor is labeled as not available.

Monitor Testing Criteria
- Engine coolant temperature must be above 170 degrees F.
- Engine has to be running for a predefined amount of time.
- Throttle position must be within a specified range.
- Manifold pressure must be within a specified range.
- Engine speed must be within a specified range.
- Airflow (measured or calculated) must be within a specified range.
- Fueling correction must be within a specified range.
- Vehicle speed must be within specified range.

Monitor Modes of Operation
- Normal should be achieved if the testing criteria is met.
- Pending is assigned if any DTCs are set for engine misfires, oxygen sensor-related faults, fuel system monitor has faulted, EVAP monitor is running, or catalyst monitor is running.
- Conflict is assigned if the ECM is in open loop, pending DTCs are set for misfires or vehicle speed sensor, crankshaft position sensor has faulted (engine speed), or oxygen-sensor-related faults.

- Suspending occurs until the oxygen sensor monitor has been completed with passing results.

Comprehensive Component Monitor

The "overlord" of the ECM's monitors is the CCM. The ECM is constantly receiving data from a variety of sensors and systems and it must determine if the data is valid. The ECM compares data from a variety of interrelated sensors and makes sure that all of their readings agree. For example, if the throttle position sensor reports the throttle as being open to 100 percent, and the manifold pressure stays at an idle reading, yet the engine speed isn't changing, then the ECM can assume that the TPS is not sending valid data, even though it hasn't set a specific throttle position DTC. The CCM makes sure that sensor data is returning in a valid amount of time. After all, even correct sensor data isn't of any value if it doesn't arrive in time.

The unfortunate thing about the CCM is that each manufacturer has established its own standards for what it tests and when it tests it; so it is very difficult to establish a standard. Therefore, you have to refer to the technician's manual for each particular vehicle when diagnosing a CCM fault.

CHAPTER 9

FOUR-STROKE ENGINE CYCLE

Suck, squeeze, bang, blow. Those four words describe the internal combustion four-stroke engine's operation. More technically, this engine operation is known as the Otto Cycle. It is named after its inventor, Nikolaus Otto, who introduced the modern-day four-cycle engine in 1876. The premise for the engine hasn't significantly changed.

In the simplest terms, an engine is nothing more than an air pump. Air is brought into the pump, work is performed, and the air is exhausted out of the pump. The larger the pump or the more air that can be moved through the pump, the more work that the pump can perform. The pump in the vehicle is the engine, and the work performed is the power generated to rotate the wheels and propel the vehicle forward. Air is brought into the engine through the intake, the crankshaft and pistons perform work, and the spent air/fuel charge exits through the exhaust system.

The four-cycle engine expands upon that simple pump example. The engine starts by drawing in fresh air (suck). The piston's movement traveling up the bore compresses this air charge (squeeze). A spark plug ignites the compressed air/fuel charge (bang), which in turn pushes the piston back down the bore creating work (torque). Finally, the piston moves up the bore, the exhaust valve opens, and the burned air/fuel charge is expelled (blow). Then, the process begins again. Manipulating performance, economy, and emissions is as simple as manipulating the air/fuel charge flow through all four cycles.

For a four-stroke engine, key parts include the intake valve, the exhaust valve, the spark plug, and the piston/rod combination. (Other engine components are not illustrated for the purpose of this example.)

Intake Cycle

The intake cycle on the four-cycle Otto engine model consists of drawing air into the chamber and mixing it with an accelerant either before or during the filling of the combustion chamber. In a naturally aspirated engine, the air is drawn into the combustion chamber because a vacuum is generated by the downward action of the piston in the bore. Nature despises a vacuum and always attempts to equalize pressure between two areas. The MAP sensor can detect this vacuum signal. The throttle blade controls the engine's power and, conversely, the engine's speed. As the intake valve opens, the air sitting in the intake manifold, which is at atmospheric pressure, rushes into the combustion chamber in an attempt to equalize the air pressure in both places.

The engine is going to demand the same volume of air each time the pistons go from top dead center (TDC) to bottom

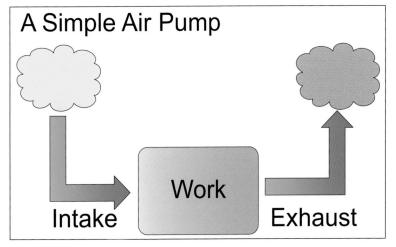

In the simplest of terms, the engine is expressed in three phases: the intake phase, the work phase, and the exhaust phase.

AUTOMOTIVE DIAGNOSTIC SYSTEMS

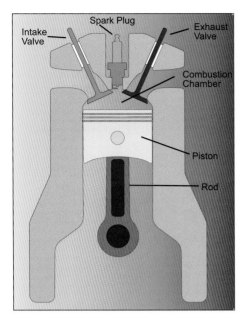

Figure 9.1. These engine components are integral to the workings of the four cycles and the Otto engine.

dead center (BDC). This volume is determined by the diameter of the cylinder bore and the length of the stroke from the crankshaft. The job of the throttle body is to moderate the amount of airflow available to the engine. Thus, the vacuum that is measured is always immediately after the throttle blade.

All vehicles equipped with OBD-II employ some type of fuel injection, so you can ignore the carburetor's role in the internal combustion engine. As the incoming air is rushing into the intake, liquid or gaseous fuel is introduced. The fuel can be added prior to the intake manifold at the throttle body in what is termed a wet manifold. The fuel is mixed with the incoming air, and the atomized air/fuel charge travels through the intake manifold into the chamber. Fuel can also be injected directly into the cylinder head after the intake manifold. This style of manifold is known as a dry manifold because it only supports air and no fuel flowing through it. The mixture's motion within the chamber ensures adequate air/fuel blending.

Once the air/fuel mixture has been added to the combustion chamber, the intake valve closes, and the piston begins to travel up the bore to start the next cycle.

Compression Cycle

As the piston rises up its bore toward the cylinder head, the volatile mixture of gasoline and air in the chamber is reduced from its initial volume at BDC to its minimal volume at TDC. This volume reduction is known as the static compression ratio. The engine's compression ratio is simply defined as the ratio of the air/fuel change in the volume in the bore from when the piston is at BDC to TDC. For example, a compression ratio of 9:1 defines that the volume of the air/fuel charge in the bore at BDC is compressed into a volume nine times smaller by the time the piston gets to TDC.

As the air/fuel charge is compressed into a smaller volume, the air/fuel charge does not transfer heat under compression. As the charge is compressed, the temperature of the charge increases without any heat being transferred into the charge from an external source. As the volume decreases and the pressure of the charge increases, the amount of energy available to be converted to work increases. From an engine-performance standpoint, the greater the air/fuel charge is compressed, the more work is available from the charge. As a result, the compression ratio is increased to raise horsepower and torque on many high-performance engines.

Also, as the air/fuel charge is compressed, there is a significant amount of turbulence and mixture motion. This helps to create a better fuel droplet distribution and evaporation throughout the air/fuel charge, which ensures a more even and controlled burn and allows for more energy or work to be produced.

There is a limit to the more-is-better theory. Continual increases in compression ratios cause ill effects, including excessively high cylinder pressures, heat and ring package failure, and detonation. Fortunately, cylinder chamber design, spark plug placement, valve placement, and piston shape can all affect the mixture, motion, and burn rates in the cylinder. Also, static cylinder compression ratios can be manipulated further by valve events to reduce the dynamic compression ratio, in order to make higher static ratios friendly to consumer-grade gasoline. As compression ratios are increased, there is an increase in nitrogen oxide, a volatile emissions gas.

Combustion Cycle

At this point in the Otto engine model, the air/fuel mixture is now squeezed into a very small volume between the cylinder head and the top of the piston. On a gasoline engine, a

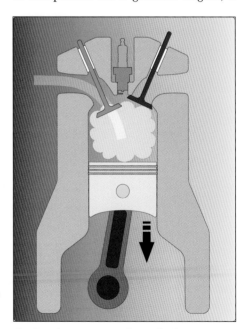

As the piston moves down the bore, a vacuum is formed. The intake valve is opened, drawing the intake charge into the combustion chamber and filling the area with a volatile air/fuel charge.

spark starts the burn of the volatile mixture. The ECM's command to initiate the spark is based on the rotational position of the crankshaft. The ECM views the engine's rotation in terms of its relationship to TDC and BDC of the piston's position in the bore. A crank position sensor typically reads the position of the crankshaft. This position is relayed to the ECM.

Based on the position of the crankshaft, the ECM determines the optimal time to fire the spark plug. This firing time is based on engine speed and vehicle load and is typically derived from a lookup table accessed by the ECM. The ignition coil is used to provide the high voltage required to fire the spark plug.

This coil can be shared among all of the cylinders, but is very rare in new vehicles. More likely, the vehicle may be equipped with an individual coil per cylinder (known as coil-on-plug), or a combination of multiple cylinders per coil (known as wasted spark).

The job of the coil is to convert the vehicle's 12 volts from the electrical system to 20,000 to 100,000 volts, which is needed to ignite the spark plug. The ECM commands the coil's magnetic field to collapse, which results in a sudden spike of voltage from the coil to the spark plug.

This high-voltage signal travels from the coil through a wire to the spark plug, and a spark that jumps across the gap at the tip of the spark plug is created. This spark then ignites the air/fuel mixture and heat and pressure within the cylinder rapidly begin to rise. The resultant pressure then pushes down on the piston, which rotates the crank and the engine performs work.

Exhaust Cycle

As the piston reaches BDC, all of the combustibles within the air/fuel charge should be mostly expended, and the cylinder now contains a mixture of mostly spent, heated gases. As the piston begins to travel back toward TDC, the valvetrain opens the exhaust valve and the spent gases are pushed out of the cylinder, through the cylinder head, and out through the exhaust system. The cylinder is now mostly void of spent gasses and is ready to draw in a fresh air/fuel charge and repeat the entire cycle.

Otto Cycle Pressure versus Volume

Even though the Otto engine model is described as four cycles, two sub-cycles exist within the main cycles. These two sub-cycles are the power stroke and heat rejection cycles. To show where these sub-cycles occur, the entire Otto cycle can be mapped in a pressure-volume diagram for the different stages. Figure 9.2 shows the relationship of pressure in the cylinder versus the volume of the air/fuel charge within the cylinder.

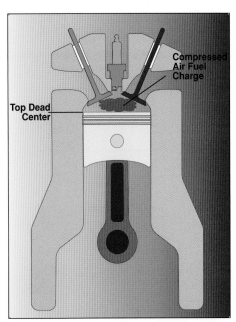

The piston moves up the bore toward the cylinder head. The intake charge is compressed into a much tighter and denser charge when the piston gets to top dead center.

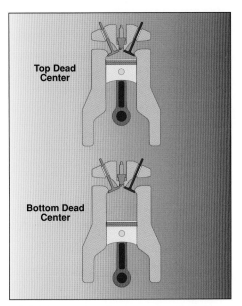

TDC is located at the upper-most position of the piston within the bore. BDC is located at the lowest position of the piston in the bore. In terms of crankshaft rotation, bottom dead center is 180 degrees from top dead center.

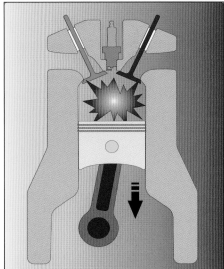

The ECM commands the coil to collapse its field, producing a high-voltage surge that causes an arc to jump across the spark plug gap and ignite the air/fuel charge in the cylinder. The ECM controls the timing of this event based on a number of degrees before or after top dead center.

CHAPTER 9

The intake cycle starts with the charge at atmospheric pressure (the least pressure on the graph), and the volume of the air charge is also at its lowest. As the piston draws downward in the bore, with the intake valve open, the pressure remains constant but the volume of the charge grows (the line moves to the right on the graph).

The next cycle is the compression cycle. As the piston moves up the bore, the pressure in the cylinder increases as the volume decreases. Temperature in the bore also rises as the charge is compressed. Work is now being performed on the air charge as the piston moves from BDC to TDC.

The combustion cycle follows (a vertical line on the graph located at TDC). When the charge is ignited, there is initially a significant increase of cylinder pressure and temperature due to the heat released by the ignition, but the volume stays constant.

The piston drives downward in the power stroke sub-cycle as the air charge burns. The pressure in the bore drops rapidly while the volume of the charge increases dramatically. Work is now being done on the piston to rotate the crankshaft.

When the piston reaches BDC, the heat-rejection sub-cycle occurs. At this point, the charge is cooling because the burn has occurred, the pressure in the cylinder is decreasing, and the volume is staying somewhat constant. Also at this point, the exhaust valve is starting to open and the spent charge is seeking equilibrium with the outside atmospheric pressure.

The final stage is shown as the exhaust cycle on the graph. The pressure is back to its lowest and near atmospheric pressure. The volume is reduced as the piston pushes the spent charge out the opening created by the exhaust valve.

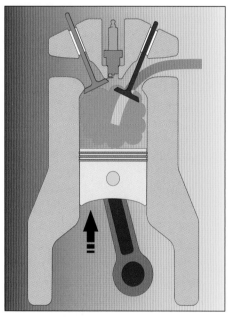

The exhaust valve is opened and the upward motion of the piston through the bore pushes out the spent charge.

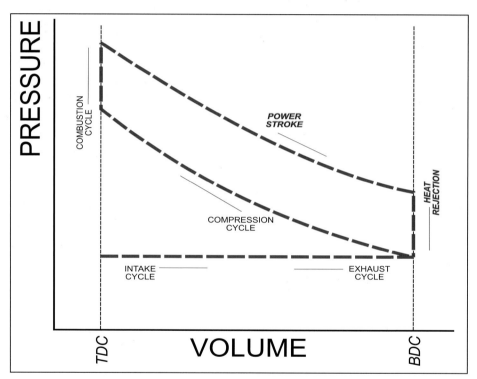

Figure 9.2. The relationship of pressure versus volume for the Otto engine is mapped to show the relationship between the two during the various cycles.

CHAPTER 10

OBD-II AND THE OTTO ENGINE MODEL

It is important to understand the operations of the internal combustion engine. Much of the health of an engine and, more importantly, its effect on emissions, fuel economy, and performance hinge upon a few basic premises: Seal the holes, flow the fuel, and burn the air/fuel charge properly and efficiently. As discussed earlier, the four steps of suck, squeeze, bang, and blow all work in harmony to achieve the emissions, fuel economy, and performance criteria.

With an understanding of the basics of the engine model, the mechanic can infer that the engine has a sealing issue, whether it's valves, piston rings, or intake gaskets. It is clear through studying and understanding the Otto engine model that the vacuum is created during the induction or "suck" stage. The downward travel of the piston in the piston bore creates a measured vacuum, and that vacuum signal travels through the intake manifold, where the pressure is measured.

Pressure and Vacuum

Atmospheric pressure is the force caused by the weight of the atmosphere resting on an object over a unit of measurement. If you could slice a 1-inch by 1-inch piece of the atmosphere that extended all the way to the edge of our atmosphere and then measure its force, it would measure roughly 14.7 foot pounds of force. The pressure near sea level is averaged to 29.21 inches of mercury. In most automotive applications, the measurement is kPa, and the pressure is 101.325 kPa at sea level. As the elevation increases, this value decreases slightly. Most automobiles recognize anywhere between 99 kPa and 101 kPa as atmospheric pressure.

Old-school mechanics generally have a vacuum gauge hanging somewhere in the shop. This tool was typically hooked up to a vacuum line on the carburetor or intake manifold and told a lot about the operating conditions of the vehicle through measuring the vacuum in the intake tract. OBD-II–equipped vehicles generally work in terms of manifold pressure reported from the MAP sensor. Although it is labeled as pressure, it actually measures pressures that are below atmospheric pressure (99 to 101 kPa). As an engine

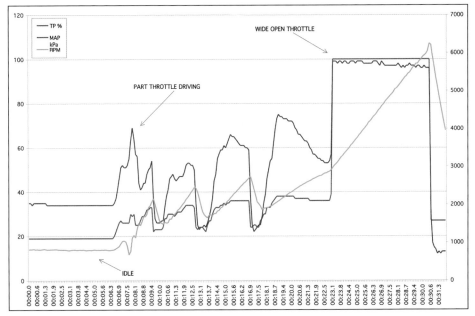

The relationship between the percentage of opening of the throttle and the pressure inside the manifold is mapped. As the throttle is opened, the absolute pressure inside the manifold rises.

AUTOMOTIVE DIAGNOSTIC SYSTEMS

CHAPTER 10

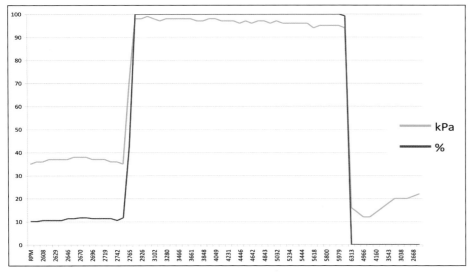

This example shows an engine on an engine dynamometer being tested for a wide-open-throttle (WOT) run. As the engine is held at a steady speed of roughly 2,500 rpm, the MAP sensor is reading around 38 kPa with the throttle being held open at roughly 11 percent. After the throttle is opened completely to 100 percent (no restriction), the MAP sensor reads 96+ kPa, indicating that there exists only a slight amount of vacuum at the throttle body, which is acceptable. A slight decrease in MAP sensor readings is seen as the engine speed climbs, which could indicate a slight restriction in the intake tract or even a dirty air filter. When the throttle body is completely closed after the dynamometer pull (back to 0 percent), the MAP sensor returns to roughly 25 kPa at idle, which is expected because a vacuum is now being generated behind the closed throttle body.

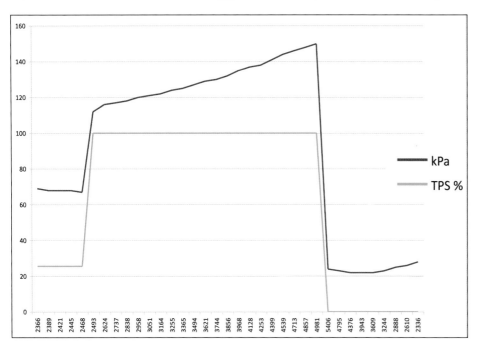

This vehicle is equipped with a supercharger. As seen on this WOT dynamometer test, the manifold absolute pressure sensor shows when the engine enters a boosted state (as labeled on the graph). After the MAP sensor reads above atmospheric pressure (100 kPa), the supercharger or turbocharger pressurizes the air/fuel charge.

runs and generates the vacuum signals in the intake track, the MAP sensor measures the pressure that is below atmospheric pressure. A typical engine idles anywhere between 28 and 40 kPa with no load on the engine.

The readings on a vacuum gauge are just the opposite of those from the MAP sensor. An engine idling at, say, 30 kPa on the MAP sensor reads on a vacuum gauge at 21.06 inHg (inches of mercury). If the throttle is opened to wide open while the engine is running, the MAP sensor reads 9 kPa while the vacuum gauge reads approximately 3 in Hg. Simply remember the inverse relationship: If the MAP sensor is rising in numbers, the vacuum gauge falls, and vice versa.

Supercharging/Turbocharging versus Naturally Aspirated

Atmospheric pressure is typically around 100 kPa (or one atmosphere). When properly functioning, on naturally aspirated engines operating at wide-open-throttle (WOT) conditions, the MAP sensor typically reads very close to atmospheric pressure minus a few kPa.

When an engine is equipped with a supercharger or a turbocharger, the air charge is then forced into the intake manifold at pressures greater than 100 kPa. Instead of relying purely on the compression of the piston to decrease the volume of the air/fuel charge, the supercharger is pre-compressing the air charge while it is filling the cylinder.

The increased air supply requires more fuel, which allows greater work to be provided by the same cubic volume of cylinder. Unfortunately, as the compression of the air/fuel charge increases, so does the temperature of the air charge, and this increase in temperature makes the air/fuel charge more prone to detonation.

Diesel Engine Architecture

The Diesel engine operates a bit differently than the gasoline engine. German inventor Rudolf Diesel invented an engine in 1878 in an attempt to improve the efficiency of the gasoline engine of the day. In 1892, a formal patent was issued for Diesel's invention.

The main difference between the two engines exists in the combustion cycle. This difference is the way each engine handles the ignition of the air/fuel mixture that propels the piston downward during the cycle. In the gasoline engine, the air and fuel are mixed into a homogenous charge, and then that charge is compressed before the spark plug is fired. The diesel engine does not use a spark plug to ignite the air/fuel charge. Instead, the air charge is brought into the cylinder and compressed without the presence of fuel. Thermodynamics shows that as the air charge compresses, heat is developed in the cylinder. This now-heated and compressed air charge has fuel introduced to it near TDC of the piston. When the atomized fuel hits the hot, compressed air charge, it ignites the fuel and starts the combustion cycle.

Unlike a gasoline engine, a Diesel engine refers to timing in terms of when the injector fires fuel into the heated air charge in the cylinder. When looking at scan-tool data, the timing PIDs refer to the timing of the firing of the fuel injector and not a spark plug timing as shown on a traditional gasoline engine. Furthermore, the amount of fuel injected is dependent on a mapping within the ECM, based on engine speed and a calculation of fuel delivery demand. This also shows up on a scan tool and is very handy information when diagnosing a Diesel engine.

Since heat is required to self-ignite the fuel, efficient combustion of the fuel charge is problematic when the air temperature or engine is too cool. In that case, the engine may experience a significant lack of combustion, which reduces efficiency and increases emissions. To address these issues, engine manufacturers employ a variety of techniques, including a glow plug. A glow plug acts like a spark plug to ignite the mixture until the heat is sufficient in the chamber, pre-ignition chambers, diverter valves, etc. All of these things make the Diesel engine much easier to use in a daily-commute lifestyle.

The main question usually asked is, "Why run a Diesel?" The simple answer is efficiency. That's what Rudolf Diesel was attempting to prove more than 100 years ago. The reason for the efficiency can be expressed simply as high thermal efficiency due to increased compression ratio, and the fact that the amount of energy available in a gallon of Diesel fuel is 129,800 British Thermal Units (BTUs) versus the 114,000 BTUs available in a gallon of gasoline.

Diesel engine injector timing map is shown with the piston at top dead center.

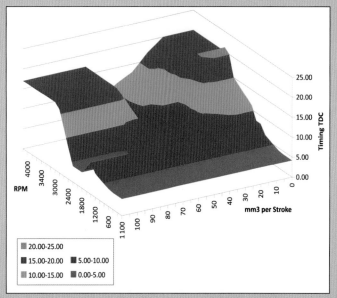

This is the Diesel engine injector timing map for quantity of fuel added with respect to manifold pressure.

CHAPTER 10

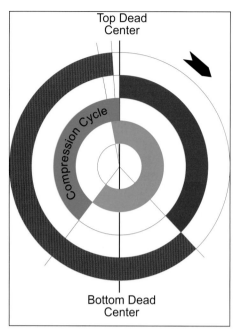

This diagram shows the amount of crankshaft rotation in each of the four cycles of the Otto engine model. Camshaft event timing dictates when each of the intake and exhaust cycles starts and ends.

Crank, Camshaft and Valves

In the standard four-cycle engine model, the crankshaft makes two revolutions to the camshaft's one revolution. Thus, it requires two complete revolutions of the crankshaft for the piston to complete its four-cycle process. The crankshaft's rotation is tied directly to the camshaft's rotation by a belt or a chain. The drives attached to the crankshaft and camshaft account for the 2:1 difference in speed.

The camshaft is responsible for the timing of when the intake and exhaust valves open with respect to the crankshaft, i.e., two of the four cycles of the Otto engine model. A camshaft's specifications detail the exact timing of these valve events in relation to the position of the piston in the bore.

For example, a camshaft may specify that the intake valve opens at 12 degrees BTDC and closes at 32 degrees ABDC. Camshaft designers can manipulate the timing of the valve events to have dramatic results on an engine's power, economy, emissions, and sound. The camshaft events can also be used to move the power band of the engine to suit the application of the vehicle. For example, a camshaft for a towing vehicle can be designed to increase low-end torque at the sacrifice of high-RPM power.

Static versus Dynamic Compression Ratios

Ever since the first mechanic hot rodded an engine, compression ratios have always been hyped. Compression ratios that most people quote are the static compression ratios—a theoretical ratio, which is purely computed. It is the mathematical representation of the ratio of the cylinder volume compressed between the piston at BDC and the piston at TDC. It is easy to calculate because all of the variables in the equation remain the same (static) except for the position of the piston in the bore. The assumption of static compression ratio is that the intake valve has no effect on compression ratio.

As discussed earlier, the engine is anything but static. The valve event for the intake cycle typically closes after BDC. Thus, the idea of the static compression ratio measuring from BDC to TDC is incorrect. Instead, a second compression ratio called the dynamic compression ratio is calculated. Because the intake valve is closed after BDC, the effective overall volume of the cylinder is decreased because the piston has risen in the bore with the valve still open. As the piston is raised, it actually forces air/fuel charge back out the intake valve until it closes completely. After the intake valve is fully closed, the starting point for the calculation of the dynamic compression ratio can be determined. So the dynamic compression ratio is always less than the static compression ratio, unless the cam is ground such that the intake valve opens and closes precisely at TDC and BDC. (I've never seen this.)

This means that the engine designer can manipulate valve events to obtain the optimal dynamic compression ratio for the vehicle and its intended use. For example, the valve events can be manipulated by camshaft lobe selection. A successful cam choice can allow optimal performance while remaining detonation friendly and running standard pump-grade gasoline.

CHAPTER 11

CONTROLLING FUEL SYSTEMS

The advent of OBD-I and OBD-II systems and advancements in fuel delivery technology (including fuel injectors and higher-pressure fuel systems) has led to significant improvements in power, economy, longevity, and emissions. Controlling these fuel components is at the heart of the ECM and the variety of sensors that provide control and feedback to the ECM. As discussed in previous chapters, the basic internal combustion engine hasn't fundamentally changed since its invention. What has changed with the internal combustion engine is the accuracy and efficiency in the controls that allow it to run. One of the largest contributors to this accuracy and efficiency is the improvements for fuel delivery and for monitoring the fuel system.

Closed Loop is the Key

In Chapter 1, the concept of a closed-loop system was introduced. The three basic aspects of the closed-loop system are action, measurement, and correction. This holds true from the simple thermostat example to the complex systems in an internal combustion engine. The ECM is responsible for interpreting incoming sensor data, controlling the fueling action, polling sensors for their measurements resulting from the fueling control, and then correcting the fuel mixture based on the information from the post-combustion sensors.

To gain an understanding of the closed-loop system of the internal combustion engine, the Otto engine model should be studied. It states that the four-cycle internal combustion engine works in four phases: intake, compression, expansion, and expulsion. From a fueling viewpoint, the ECM is concerned with the first and last phases, the intake and the expulsion. The ECM has no influence on the compression phase, which is mechanically determined. The ECM does not determine the expansion phase per se; rather it is largely determined by the timing driven by the ECM. (This is discussed in Chapter 13.)

In the most basic closed-loop system for the Otto model, the ECM is concerned with four things:

1. Measure incoming air
2. Add fuel based on incoming air and engine load
3. Observe results of expelled exhaust from engine
4. Correct fuel amount

The simplest description of any engine is that it is an air pump. The more

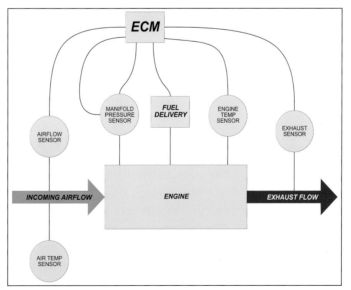

A simplified flowchart of the fueling control model. The ECM needs to measure the incoming airflow, add the required fuel to match that airflow, observe the resultant air/fuel ratio, and finally make corrections to the fuel for the next go around. This is the basis of how the ECM treats the fueling model.

CHAPTER 11

Open-Loop Systems

Many times an ECM runs the engine in what is known as open-loop mode and, as a result, the feedback portion of a closed-loop system is removed. The engine is still taking device readings from sensors, such as the engine speed, incoming air temperature, manifold pressure, etc. But the ECM is not sampling the exhaust and dynamically adapting the fueling model to suit the situation. Open-loop mode can occur at a variety of times including:

- Cold starts
- Hot starts
- Incapacitated or damaged sensor (also known as limp-home mode)
- WOT operation
- Extended idle times

The main issue is the time it takes to heat the exhaust system and its feedback sensors in order to gain accurate data. This reduces fuel economy and increases emissions, compared to closed-loop operations. Fortunately, the ECM does not stay in an open-loop state for very long. For the fueling state, most scan tools inform that the ECM is using PID 0x03. It is important to verify that an ECM is going into closed-loop mode at some time during the engine's operating cycle. Failure to enter into closed-loop mode during normal operations typically requires a diagnostic of the key sensors.

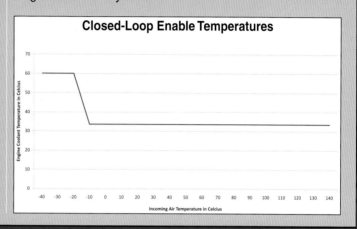

power the engine makes, the more air that the engine has to move to accomplish it. Therefore, if the amount of air moving through the engine is graphed versus RPM, it looks similar to Figure 11.1.

Increased load on an engine requires more power and, thus, more air. Therefore, the amount of air flowing through the engine is dynamic based on both engine speed and load. As seen in Figure 11.2, engine speed (RPM) is not the sole factor for the volume of airflow. Many times, the same RPM may have a different engine load based on multiple reasons, including road grade, heavily loaded vehicle, altitude variation, and other factors. The ECM must take into account not only airflow but also the engine load when calculating a fueling solution. (The ECM now has three input variables that it must determine in order to move forward with the fueling requirements.) After the incoming airflow, engine load, and engine speed are determined, the ECM can determine the precise amount of fuel to add to the engine.

The ECM must precisely control the exact amount of fuel required for the situation. In early ECM logic, variables were looked up on a table and a fueling solution was established. On later-model vehicles, high-order mathematical polynomial equations are used to dynamically calculate the precise requirement of fueling. (This book is sticking with the lookup model because it is much easier and clearer to understand.)

Modern-day engines employ fuel injectors to deliver the fuel in an accurate and timely manner. In order to control the injectors, the ECM calculates exactly how much fuel the injectors flow in a certain number of milliseconds when the injector is opened. The ECM commands each injector to open a certain amount of time to deliver a precise amount of fuel. As a result, the ECM squirts a given amount of fuel into each cylinder at the proper time, and this is based on a required air/fuel ratio and the determined amount of air entering the engine.

So in the closed-loop system, the ECM analyzes the incoming air, load, and RPM, and squirts the predetermined amount of fuel into the engine for the particular set of conditions. The ECM then needs to monitor the gases coming out of the engine during the exhaust cycle in the Otto engine model. The ECM is looking to see if the current fueling solution contains excess fuel or not enough fuel. By having a sensor sitting in the exhaust stream, it can sample the fuel left over from the combustion process and report its findings to the ECM. The feedback system flow chart now becomes a bit more complicated as shown in Figure 11.3. The complete feedback system

CONTROLLING FUEL SYSTEMS

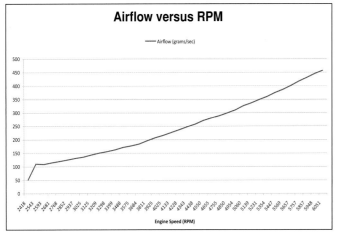

Figure 11.1. As the engine speed increases in RPM, the airflow through the engine also increases. Since the engine is fundamentally an air pump, the greater the speed of the pistons, the more airflow (measured in grams/sec) is seen going through the engine.

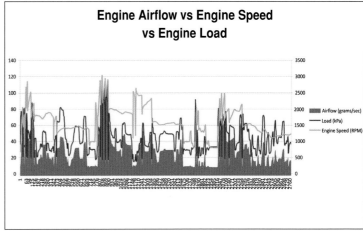

Figure 11.2. A mapped correspondence exists between the physical load on the engine and the required engine airflow. Typically, as a greater load is placed on the engine, the engine requires more airflow. Remember that, as an air pump, the more air an engine ingests the more power it typically creates.

contains three different measurement sensors that give readings to the ECM, so it can intelligently determine the proper fueling maps to run the engine at peak performance and efficiency.

Fuel Combustion and Thermal Efficiency

On the Otto engine model, the first cycle is the intake of fresh air and fuel. This mixture is compressed into a volatile mixture primed for explosion. The key question from a mathematical and essentially an ECM viewpoint is how to calculate what is occurring in the combustion cycle and determine the proper mixture of incoming air and additive fuel.

Gasoline contains energy that is released during the combustion process. A gallon of gasoline weighs roughly 6.25 pounds and contains approximately 17,500 BTUs per pound. Converting this number to gallons gives the standard acceptable value of 114,000 BTU per gallon.

The next thing to figure out is the relationship of BTU to horsepower generated. The standard conversion is 1 horsepower is equivalent to 42.4 BTUs per minute.

So if an engine at WOT consumes 25 gallons of gasoline per hour in a laboratory testing station, the horsepower derived

The Carburetor

Before modern fuel-injection systems existed, the carburetor was the only means of supplying fuel to the engine. The carburetor, as perfected as it was over many decades, is still an imprecise method of delivering fuel to the engine. In the most simplistic terms, the carburetor is nothing more than a cylindrical tube, with a smaller section in the middle. This necked-down area is known as the venturi, and the function of its smaller diameter is to reduce the pressure of the incoming air in the carburetor. A fueling orifice in the venturi area is directly tied to the fuel reservoir in the carburetor. Nature always tries to balance pressures to achieve equilibrium. Because the pressure inside the fuel reservoir is greater than that in the venturi, it pushes air and fuel out the orifice and into the lower-pressure area of the carburetor.

Unfortunately, this isn't a dynamic system, at least not dynamic enough to respond to the fueling considerations required in a modern-day-emissions and CAFE engine design. The designers added more tuning circuits in the carburetor's meter blocks to achieve better fuel delivery based on the current operating conditions. But in the end, it just wasn't precise enough at delivering fuel economy and retaining emissions quality. So the carburetor design was basically phased out of the modern automotive system.

CHAPTER 11

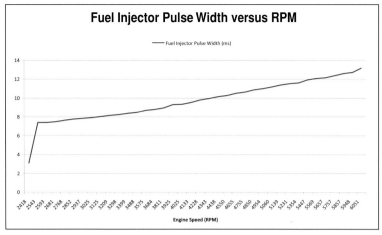

injector is open, which increases the fuel added to the engine. More power means more fuel consumption.

As engine speed increases, so does airflow and the load on the engine. The engine then requires more fuel to run. The ECM controls the time that the injector is open, which increases the fuel added to the engine.

from this measurement is expressed as:

25 gallons of gasoline in an hour x
6.25 pounds per gallon =
156.25 gallons per hour

Given that 1 gallon of gasoline provides an energy equivalent of 114,000 BTU, the hour-long test at WOT of this engine shows that the engine has created:

156.25 pounds per hour x 17,500 BTU per pound = 2,734,375 BTU per hour

Converting that to horsepower, the total horsepower that the engine is outputting calculates to:

2,734,375 BTU per hour ÷ 2,542 BTU per hour in 1 hp = 1,075 hp

But the test engine is a 346-ci engine, factory-rated at 305 hp. Why such a difference between 1,075 hp and 305 hp? The key is the thermal efficiency of the combustion process.

During the combustion process, thermal losses occur at many points. Energy is primarily thought of as moving the piston up and down, but some of that energy is used to heat the exhaust gases, heat the engine block, heat the oil, etc. Not all of the energy is available to move the piston. So it's important to understand the thermal efficiency of the engine, which can be expressed by:

Horsepower = Thermal Efficiency x
Fuel Flow (in pounds per hour) x BTU per pound of fuel ÷
BTU per hour per horsepower

Given the known factors, the equation becomes:

Horsepower = Thermal Efficiency x
Fuel Flow x 17,500 ÷ 2,542

Or:

Horsepower = Thermal Efficiency x
Fuel Flow x 6.8843

Since we are interested in thermal efficiency, the equation can be rewritten as:

Thermal Efficiency = Horsepower x
0.1453 ÷ Fuel Flow (in pounds per hour)

In the test of the 305-hp engine earlier, it was determined that the hour of running at WOT used 156.25 pounds of gasoline. Thus, thermal efficiency can finally be calculated as:

 Weight of Gasoline

In this book, I refer to the weight of gasoline as 6.25 pounds per gallon. Unfortunately, depending on the grade of gasoline and the additives for a specific blend and vendor preferences, these weight values can vary. For example, a standard grade of gasoline from a vendor in my home state of Missouri weighs exactly 6.26 pounds per gallon. The same vendor sells a premium-grade gasoline weighing 6.32 pounds per gallon. Changing vendors and re-sampling the two grades can result in a different set of measurements.

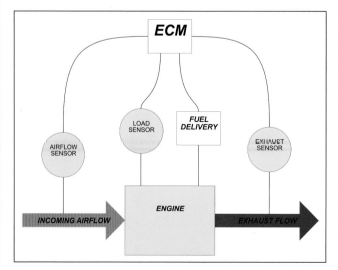

Figure 11.3. Closed-loop fuel control flowchart requires only a few sensors, including airflow engine load and exhaust feedback measurements. All of these sensors feed back into the ECM, so it can control the amount of fuel required by table lookup or fueling algorithm.

72 AUTOMOTIVE DIAGNOSTIC SYSTEMS

Alpha-N Fuel Metering

When the first ECMs appeared in vehicles in the early 1980s, the processing speed and storage availability trailed today's high-school student's calculator! As a result, the first fuel-control algorithms were very basic in their capabilities. The first such systems were known as Alpha-N. In a nutshell, only two sensors were required to determine the fueling calculations. The engine speed in RPM and the position of the throttle blade allowed the ECM to simply look up a pseudo-VE value stored on a table in the ECM. This value was then multiplied by the required air/fuel ratio to determine the amount of fuel that the engine required.

Even today, some high-performance off-road vehicles employ the Alpha-N method because it requires very few sensors, which increases their off-road reliability. Most importantly, many of these off-road engines suffer from very poor manifold vacuum due to radical, off-road camshaft designs. Because the vacuum or pressure readings in the manifold play an integral part of defining the required fuel in more modern systems, the speed-density or mass-airflow models tend to react very poorly with the low manifold vacuum readings of these style engines. The issue becomes that, with a very small range of manifold pressures, say, from 70 to 100 kPa, this lack of resolution makes it very difficult to cover the broad range of operating conditions of the engine using a speed-density or mass-airflow model. Figure A shows that the limited resolution of manifold pressure readings causes a limited range of fueling solutions, meaning the engine will not run optimally for power and economy.

By tracking the fueling changes via readings from the throttle position and engine speed, the Alpha-N fueling algorithm gives a much broader operating range capability as shown in Figure B. The downfall of the Alpha-N system is that it can't adapt to the changes in air density and load situations. So this system is simple and works for off-road vehicles and their particular situations. But for overall economy, performance, emissions, and longevity, the Alpha-N system is not a good solution.

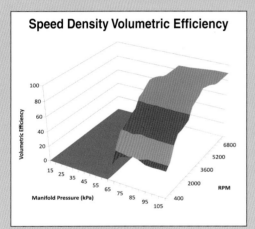

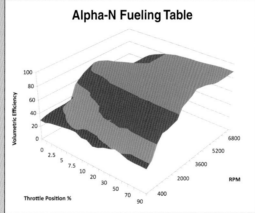

Figure B. If an engine has a poor vacuum signal or high manifold pressure, the ECM may not be able to control fueling properly. An Alpha-N fueling algorithm allows the ECM to control fueling for a wide operating range because it uses a VE table driven by throttle position and engine speed.

Figure A. Due to the lack of range of manifold pressure (or vacuum), the ECM may be working with a truncated VE table, which limits its range of fuel control. When faced with off-road engines that exhibit these tendencies, other options such as Alpha-N may be applicable.

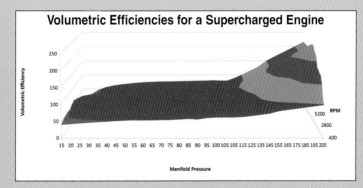

When forced-air induction is added to an engine, pressurized air enters the combustion chamber. As a result, the air exceeds atmospheric pressure, and the volumetric efficiencies must be increased to take this into account.

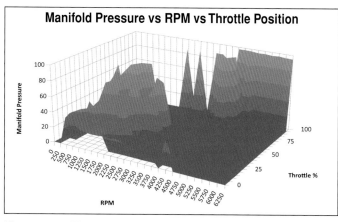

Figure 11.4. The graph of a drive cycle of a vehicle shows that the manifold pressure varies significantly. The throttle position and engine speed have a great effect on these readings. Only when the engine is at or near WOT do the manifold pressure readings approach atmospheric pressure levels.

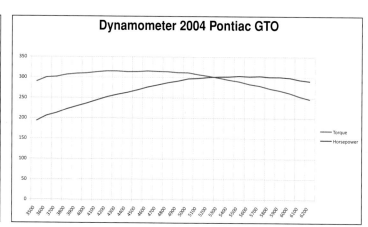

Figure 11.5. The VE chart stored in the ECM shows that the VE varies throughout the engine speed and load range. As expected, the peak volumetric efficiencies occur near the peak torque engine speeds.

Thermal Efficiency = 305 hp x 0.1453 ÷ 156.25 pounds per hour

Or:

Thermal Efficiency = .2835 or 28.35 percent

So, based on the observations of the test engine, the engine itself is very numerically inefficient—coming in at a thermal efficiency rating of 28.35 percent!

Now, given the thermal efficiency of the engine as a known entity, the equation can be rewritten to calculate the theoretic fuel requirements for creating a certain amount of horsepower:

Fuel Required (in pounds per hour) = 0.1456 x Horsepower ÷ Thermal Efficiency

Volumetric Efficiency

The Otto engine model basically describes the engine as an air-moving pump. In terms of capacity, the engine is defined by the total volume of air that it can displace. This measurement has a unit value of cubic inches or liters and is typically noted as the engine displacement. In the example, the engine is a 346-ci engine, which means that it can displace a total of 346 ci of air. Remember that the engine's crankshaft makes two complete rotations to go through the entire four cycles of the Otto engine model. Thus, one revolution of the crankshaft equals one half of the total engine displacement; in this case, 346 ÷ 2 = 173 ci of displacement for each crankshaft rotation.

Theoretically, a naturally aspirated engine should flow 100-percent volumetric efficiency (VE) when the engine is performing at peak torque levels. Unfortunately, as has been shown throughout this book, theory doesn't always manifest in the real world. If an engine operates at 100-percent VE, the pressure inside the cylinder is equal to atmospheric pressure. But the measurable vacuum signal in the intake manifold varies throughout the engine's operating RPM range. The presence of a vacuum in the intake manifold indicates that there is a vacuum inside the cylinder, and thus, the engine is not operating at 100-percent VE. Thus, the higher the vacuum reading (or the lower the manifold pressure reading) observed, the lower the volumetric efficiency is, based on the engine's current operating state.

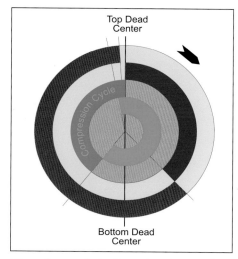

Since the crankshaft requires two revolutions to complete the four cycles in the Otto engine model. Because each piston travels up and down the bore twice, the actual displacement of airflow per crankshaft revolution is 1/2 the total cubic inches. The purple area represents one revolution of the crankshaft while the yellow area represents the second rotation of the crankshaft.

To validate VE during operating conditions, a vehicle was driven and critical information was recorded. During the drive, the data of the throttle position (in percentage), the manifold pressure

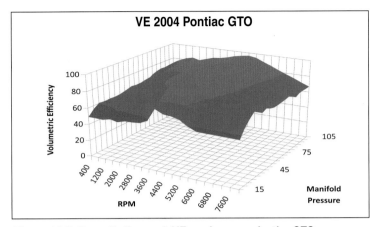

Figure 11.6. To verify the peak VE engine speeds, the GTO was placed on the dynamometer to record horsepower and torque. The engine speed for peak measured torque coincides with the peak VE table stored in the ECM.

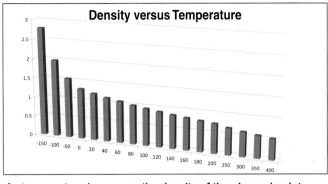

As temperature increases, the density of the air coming into the engine decreases. This is why it is important to be able to measure the incoming air temperature when calculating the mass of the air coming into the engine. Even a 20-degree C change in air temperature can result in a 5-percent difference in calculated air density.

(in kPa), and the engine speed (in RPM) was recorded. Figure 11.4 shows this mapping. Please note that the manifold readings were lower when the engine was run at part throttle and at lower speeds. This indicates a vacuum was being created in the intake manifold. As the throttle was opened and the RPM increased, manifold pressure rose and, conversely, the intake manifold vacuum decreased. Considering that the engine reached manifold pressures very close to 100 kPa (atmospheric pressure), the readings indicate that the engine was operating near 100-percent VE.

Peeking inside the ECM yields the derived VE maps for this particular engine. The ECM tracks the VE based on the readings of the speed of the engine in RPM and the manifold pressure readings in kPa.

In theory, the engine should reach peak torque when it reaches its highest VE. It is possible for a naturally aspirated engine to reach 100-percent VE, but it is difficult, takes specialized equipment, and is very expensive. Typically, a very efficient engine runs around 95-percent VE. The higher the VE of a particular engine, the more power that the engine can produce. Figure 11.5 shows the mapping of the VE and shows the peak VE readings occur at near-atmospheric pressures in the manifold and at engine speeds between 4,400 and 4,800 rpm. This particular vehicle was then placed on a chassis dynamometer and the horsepower and torque were measured at WOT.

Data from the dynamometer is plotted in Figure 11.6. The dynamometer measured peak torque at the drive wheels at an engine speed of 4,600 rpm. This correlates to the VE maps in the ECM itself, and proves the theory that the peak VE is at the peak torque readings for the engine.

Airflow Volume

After you understand and calculate the efficiency of the engine, the next thing to understand is the overall amount of air flowing through it. Once again, in theory, the amount of air drawn through the intake is constant regardless of engine speed or engine load. In application this volume of air changes a little bit based on the dynamics of the camshaft and exhaust systems, but this amount of change is so small that it can be disregarded for this discussion.

Given that the ECM now understands the relationship of VE versus load and engine speed, it can now calculate the volume of airflow at any point along the engine's operating parameters. Why do this? Because air density varies due to elevation changes (pressure) and temperature changes. That makes it important to level the playing field, so that the ECM can calculate the volume of the air coming in and not its mass.

Here's how the ECM calculates the volume of the airflow coming in: The ECM has its own sampling from which to determine the current air/fuel ratio. A sensor sits in the exhaust stream and allows the ECM to precisely monitor the air/fuel ratio in real time. The ECM closes the loop in terms of fueling feedback. It can analyze how far the current exhaust stream is from the desired air/fuel ratio. The ECM can then dynamically manipulate the fueling table, so that the engine continues to operate at the proper air/fuel ratio for the situation.

What Can Go Wrong?

As with any system, you need to understand what each component provides and what occurs if single or multiple components produce the wrong results or fail completely. The closed-loop fueling system that has been laid

CHAPTER 11

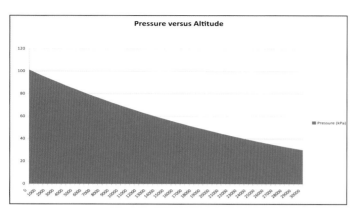

When the altitude of the operating environment changes, the atmospheric pressure also changes. Since the pressure is used to determine loading and fueling, it is important to be able measure this altitude change.

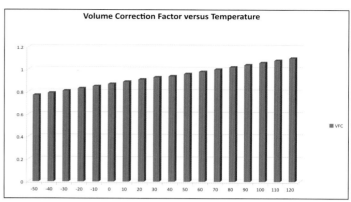

The temperature of the incoming air has a significant effect on the volume of air in the cylinder. The ECM must be able to compensate for air temperature when calculating the engine's fueling requirements.

out so far consists of the following components:

- Device to measure incoming airflow
- Device to measure incoming air temperature
- Device to measure engine speed
- Device to measure manifold pressure or vacuum
- Device to deliver required fuel
- Device to measure air/fuel ratio in the exhaust
- An ECM to control the entire system

Luckily, all of these devices can be monitored using a scan tool to determine if the readings from each device appear to be within the normal accepted range. Failure to get adequate data from a particular sensor can make the entire fueling calculations cause the engine to run rich (less than 14.7:1 air/fuel ratio) or lean (greater than 14.7:1 air/fuel ratio). At best this can hurt fuel economy and emissions. At worst, it can cause physical damage to engine components or even fail to allow the engine to run at all. Understanding what these devices should report through a scan tool, and how an incorrect reading causes problems, is the real method for diagnosing and using OBD-II tools.

Evolution of the ECM

As electronics have become more powerful and less expensive, the automotive ECM has been a significant beneficiary of these advancements. In the past, the ECMs compensated for their lack of processing speed by storing maps made up of data points that the ECM referenced by looking up one or two axes of sensor information. The sensors didn't take up any processing time, so their returned values were basically real-time data and free from a processor standpoint. This made the lookup table methodology quick and accurate.

Unfortunately, those lookup tables assume that most things stay fairly stable and have conditions that replicate the factory lab conditions where the original tables were developed. Granted, thanks to feedback systems, many of the tables can be altered to accommodate some operating differences, but it still has its limits. With faster processors in ECMs as well as more advanced sensor technology, newer fueling algorithms do not use static lookup tables.

New algorithms now describe VE calculations and fueling calculations through compound polynomial mathematical equations. This allows an extremely rapid way of calculating fueling curves. But more importantly, the system is now dynamically developed; it's based on the input of multiple sensors and feedback from other sensors. An ECM is no longer confined to a static set of data. The ECM can now adapt much quicker and more accurately to changing conditions for the engine. What started as a simple table-lookup computer has become a self-learning ECM that adapts to changing conditions throughout the entire life cycle of the vehicle.

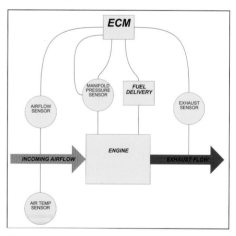

The air temperature plays a major role in determining air density. So the ECM must measure this to determine the amount of air ingested by each cylinder. The capability to measure the incoming air temperature is added to the system flowchart.

CHAPTER 12

Dynamic Fuel Correction

The ECM has the capability to accurately control the fueling of the engine based on known sensor inputs, VE calculations, fueling curves, and more. But, what occurs when the operating conditions change and the fueling isn't accurate? When this occurs, the feedback system helps to keep everything in check and balanced in the dynamic operating conditions.

Fuel Requirements

The ECM is dealing with many variables and constants when determining the fueling curves. To simplify things, the ECM looks at the variable amount of incoming air, the known constant of the fuel-injector flow rate, and the results measuring the air/fuel ratio in the post-combustion exhaust stream. This simple system allows the ECM to measure incoming airflow, determine the mass of the air, inject a known amount of fuel into the engine based on the air calculations, and then verify if the air/fuel ratio coming out of the engine is correct. This is the essence of the closed-loop fueling algorithm.

The ECM's base fueling algorithm is programmed at the factory, and it determines how much fuel to add based on the current operating parameters. The ECM has a base understanding of the volumetric efficiencies of the engine operating under ideal (laboratory) conditions. It can determine the amount of incoming air and calculate the required amount of fuel to reach its commanded air/fuel ratio. The commanded air/fuel ratio is typically a lookup table stored in a program of the ECM, and this program references engine load and engine

Figure 12.1. In order to accommodate cold-start and hot-start routines as well as calculate air densities, an additional sensor that measures the engine temperature is added. The engine temperature sensor typically measures the coolant flowing throughout the engine to eliminate the possibility of measuring a hot spot in the engine.

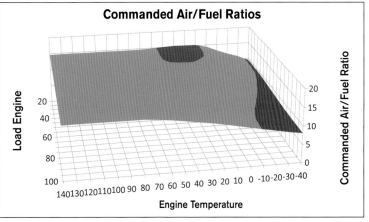

Figure 12.2. The commanded air/fuel ratios are based on the engine temperature and engine load. In order to help cold starts and cold engine temperatures prior to warm ups, the air/fuel ratio is intentionally run richer.

CHAPTER 12

Stoichiometric Air/Fuel Ratio

Stoichiometry is the basis for all chemical reactions in which a particular quantity of substances enters into a chemical reaction and produces by that same chemical reaction another quantity of substances. Typically, a ratio defines the quantities of substances going in and coming out of the reaction. In theory, the stoichiometric air/fuel ratio has enough air added to completely burn the fuel so that there is none left over after combustion.

Different fuels have different stoichiometric values, as shown in Figure A. On dual-fuel-supported vehicles, the ECM determines the percentage of each compatible fuel type and automatically changes the relationship of that percentage and required stoichiometric air/fuel ratio. For example, E85, a popular alternative fuel, is capable of running in a variety of flex-fuel vehicles. But, the stoichiometric air/fuel ratio for E85 is much lower than the stoichiometric air/fuel ratio for gasoline. Also a vehicle owner may have a quarter tank of gasoline left and then fill the remaining 75 percent of the tank with E85, which makes it a blend between the two. What the ECM must do is detect the percentage of both fuels, through a sensor in line with the fuel system or by determining the percentage post-combustion. Based on this information, the ECM adjusts the required stoichiometric air/fuel ratio for the fuel blend and adjusts the amount of fuel being added based on the fueling calculations. Figure B shows a comparison of the commanded air/fuel ratio table loaded in the ECM software for gasoline and for straight E85. The ECM must dynamically calculate the average or blending between these two tables and use that to derive the fueling calculations in real time.

The stoichiometric air/fuel ratio is typically considered a lean air/fuel ratio. This air/fuel ratio is best suited to light-load engine operation because it tends to burn much hotter in the combustion chamber and is prone to detonation when the load increases. As a result, the ECM adapts to a richer air/fuel ratio when the load or engine speed increases past a predefined point. Figure C shows that incoming air temperatures and/or engine coolant

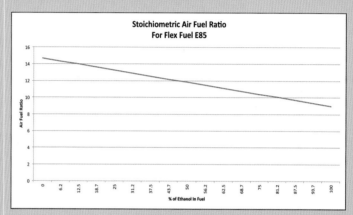

Figure B. E85 flex-fuel-equipped vehicles can mix gasoline and E85. As a result, the stoichiometric air/fuel ratio must be calculated based on the percentage of both fuels present in the fuel stream.

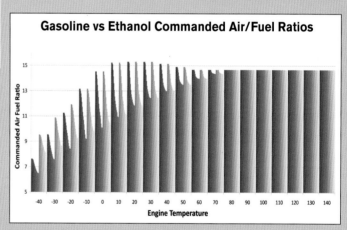

Figure C. The colder-burning E85 fuel requires a much richer mixture as the engine temperature gets colder. The ECM is responsible for looking up the proper air/fuel ratio based on the engine temperature and calculates the blended air/fuel ratio based on the makeup of the fuel stream of the blend of gasoline and E85.

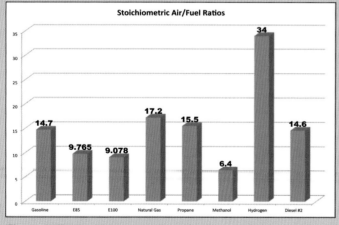

Figure A. The stoichiometric air/fuel ratio for different fuels varies depending on their chemical makeup.

temperatures can affect the air/fuel ratio. When the incoming air temperature is extremely cold, more fuel is required to operate the engine in a warm-up mode or in a closed-loop mode.

Another fueling mode, known as Lean Cruise, is available on overseas vehicles but not in the United States. Lean Cruise implements an even higher air/fuel ratio than stoichiometric in order to shave off a bit of fuel usage and while increasing mileage (Figure D). Lean cruise is not without fault though. The increase in air/fuel ratio raises emissions, particularly nitrogen oxide emissions. It also raises engine temperatures and the risk of ignition detonation under loads. Typically, lean cruise is only activated under light loads to reduce the engine damage risks, but it does not address the additional emissions that are emitted from the engine while this mode is active. In the testing performed on vehicles using lean cruise, we measured roughly 1 to 2 percent additional fuel economy, but this was on off-road-only vehicles, which were not under EPA mandates for emissions.

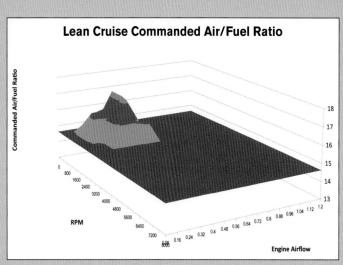

Figure D. Although not EPA-legal on vehicles in the United States, lean cruise pushes the air/fuel ratios much higher than the stoichiometric air/fuel ratio of gasoline.

temperature. Now, we've added another sensor into the flow chart, so Figure 12.1 now helps update the system flow.

Based on the sample data shown in the commanded air/fuel ratio table in Figure 12.2, and given the VE table or equation that determines the amount of incoming air, the ECM can properly calculate fuel requirements and command the fuel delivery system to very accurately meter and deliver that fuel. This metering allows the ECM to achieve the prescribed air/fuel ratio for the condition.

In a Perfect World . . .

Based on the airflow measurements and calculations, the ECM controls the amount of fuel sent into the engine. Therefore, the ECM's air/fuel ratio should perfectly match the air/fuel ratio results coming out of the engine's exhaust . . . in a perfect world. But we don't live in a perfect world, and the engine is required to operate in environments that can vary dramatically. An engine operating at the

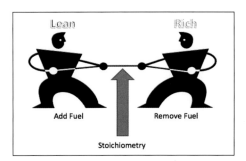

Because the engine is a dynamic system, there is a constant tug of war going on between rich and lean exhaust air/fuel ratios. The ECM balances between the two readings and adjusts accordingly.

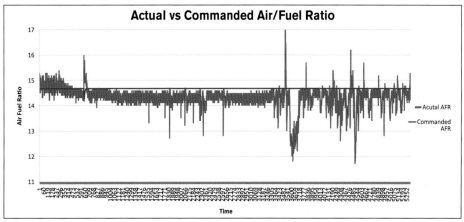

Figure 12.3. The plotting of air/fuel ratios comparing the readings from a wide band air/fuel sensor placed in the exhaust stream versus the commanded air/fuel ratio from the ECM. Note how the actual air/fuel ratio swings a lot over time and doesn't always track with the commanded air/fuel ratio.

CHAPTER 12

high altitudes of Colorado has different requirements than one operating below sea level in Louisiana. A vehicle driven in the cold winter of Minnesota has different operating parameters than one being driven through Death Valley in California. Furthermore, manufacturing tolerances of engines, tolerances in sensor readings, wear and tear on components, and any aftermarket modifications can further skew the air/fuel results from the targeted ECM-commanded air/fuel ratio. So the commanded air/fuel ratio seldom matches the actual sampled air/fuel ratio in the exhaust.

This is a key reason why the ECM and engine controls form a closed-loop system. The exhaust sensor detects the air/fuel ratio in the exhaust stream and reports its findings back to the ECM. Based on this feedback, the ECM can adjust the amount of fuel that it is commanding in order to achieve its commanded air/fuel ratio. Not only does this adjustment help in fuel economy, but it also impacts performance, engine longevity, and exhaust emissions. The question to answer is: how does the ECM make these adjustments?

Fuel-Trim Adjustments

The ECM is constantly getting feedback about the air/fuel ratio in the exhaust system. This feedback informs the ECM whether or not the actual fuel ratio is rich or lean from the commanded air/fuel ratio. For example, if the targeted air/fuel ratio is 14.62:1, and the exhaust stream is sampled and found to be 14.5:1, the ECM computes that the actual air/fuel ratio is richer than the commanded ratio. It then reduces the amount of commanded fuel in order to achieve the targeted air/fuel ratio. Conversely, if the exhaust stream sample showed an air/fuel ratio of 15.2:1, then the ECM computes that the actual air/fuel ratio is leaner than the commanded ratio and it adds to the amount of fuel that it is commanding.

The engine operating model is a dynamic one; things are constantly changing. When sampling the exhaust stream, the air/fuel ratio is constantly changing; it never stays at one ratio.

Figure 12.3 shows a graph of the actual measured air/fuel ratio using a wide band air/fuel sensor versus the ECM's commanded air/fuel ratio. Note that the commanded air/fuel ratio the ECM is targeting remains fairly constant (this engine was already up to operating temperature). But the actual air/fuel ratio that was measured in the exhaust stream varied significantly from what was commanded. Reasons for this vary and are beyond the scope of this book, but it is important to understand the air/fuel ratio is actually varying quite a bit. The key is how the ECM interprets this information and what it does with it.

Real-Time (Short-Term) Fuel-Trim Adjustments

As previously discussed, the actual air/fuel ratio is constantly changing. It hovers around the desired air/fuel ratio that the ECM is commanding, but never stays stable. As operating conditions change, the air/fuel ratio measured in the exhaust system may be slightly lean or slightly rich. The ECM is responsible for constantly trimming its fuel calculations to hit its optimal, calculated air/fuel ratio. This real-time adjustment is known by a variety of terms, such as short-term fuel trims, block integrator, etc. All these terms relate to the instantaneous fuel

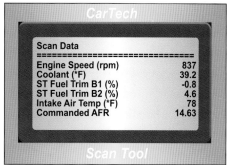

The scan tool reports the short-term fuel trims in real time. The scan tool experiences a lag and lack of data reported by the ECM because of the speed difference of the scan tool interface and the speed of the events occurring in real time. Typically, a mechanic concentrates on very high readings in the short-term fuel trims to locate a problem.

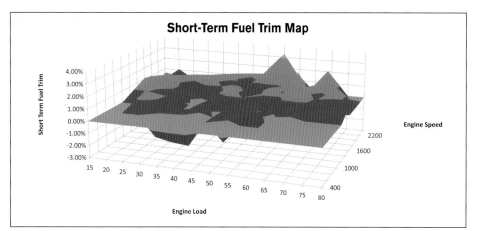

Figure 12.4. This shows a mapping of the short-term fuel trim based on the engine speed and engine load. The short-term fuel trims show the instantaneous fuel correction, which can typically vary ±4 percent. It is very rare for the short-term fuel trims to be exactly 0 percent. In this graph, areas that are shown as 0 percent are combinations of engine speed and engine loads, which were not reached in this drive cycle.

adaptation based on feedback from the air/fuel readings in the exhaust system.

The ECM constantly uses the short-term fuel trims to adjust its fueling algorithm to achieve the stoichiometric value for air/fuel stored in the ECM's list of constants. By adjusting the time at which it injects fuel into the engine through the manipulation of the injector pulse width, the ECM can attempt to add the proper amount of fuel. As shown in Fig 12.3, this is a moving target. In reality, the ECM can come very close to hitting the stoichiometric air/fuel ratio, but it really can't maintain it perfectly. The system is just too dynamic to expect that.

The ECM adjusts the fueling by monitoring the feedback from the exhaust air/fuel ratio sensor and comparing its readings to the commanded air/fuel ratio. Once determined, the ECM calculates the percentage difference from the commanded ratios to the actual ratios. It then increases or decreases the amount of fuel that it is injecting into the engine by that percentage.

For example, say the current engine load, engine speed, and incoming airflow requires an injector to be pulsed for 3.500 milliseconds to deliver the calculated air/fuel ratio of 14.62:1. If the exhaust air/fuel ratio sensor reports that the air/fuel ratio is 14.48:1, then the ECM can calculate that the ratio is about 1 percent on the rich side of the fueling curve. It can then reduce the injector pulse width by 1 percent of the standard, or it adjusts the injector pulse width from 3.500 milliseconds to 3.465 milliseconds. Keep in mind, all this is done in real time, so you can imagine the speed and processing power that it takes to keep all of this in check.

Typically, the short-term fuel trims are limited to how much trimming they can actually accommodate. Some vehicle manufacturers limit this percentage of change to ±10 percent, while other manufacturers allow the short-term fuel trims to adjust it by as much as ±25 percent. Regardless of how much it is allowed to change the fuel trims, the main goal is to keep the fueling so that it can come as close as possible to establishing the commanded air/fuel ratio in real time.

Let's examine a vehicle that a scan tool was connected to and taken out for a drive cycle. The short-term fuel trims were tracked and placed in a table that mimics the VE tables.

Figure 12.4 shows how the amount of short-term fuel trims varies significantly over a range of the engine's operating environment. In reality, the engine was operating within ±4 percent of the optimal fuel ratio, which is really pretty good for a factory engine. Why does this differ from the factory fuel settings? Again, the factory fuel tables were established in a very controlled environment and are targeted to an average setup. Due to the differences in engines and actual operating environments, the ability to feed the corrections into the fueling model is the best way to ensure that the fueling is accurate when based on the stoichiometric air/fuel ratio that the ECM is trying to achieve.

There are a couple of considerations when using a scanner to view the short-term fuel trims. First, the number of short-term fuel trims is equal to the number of banks that an engine has. So an inline four-cylinder engine typically has only one bank. A V-6 or V-8 engine has two banks. Thus, the scan tool reports the short-term fuel trims in terms of which engine bank it is trimming. Typically, a number is assigned after the acronym for the short-term fuel trim, such as STFT1, which indicates that the scan tool is reporting the short-term fuel trim for bank 1.

Another point to consider is the speed of the scan tool itself. At a mere 2,000 rpm, or 33.3 revolutions per second, a V-8 engine's crankshaft is completing a revolution every 30 milliseconds, or 60 milliseconds for an entire Otto engine cycle to complete. Imagine now the ECM reacting to readings from the exhaust air/fuel ratio sensor and making changes in that short time and then showing that short-term fuel trim reading on the scan tool. Because the data stream on the scan tool is not operating at anywhere near that speed, there is a lag and dropout rate from the actual short-term fuel trim corrections occurring in the ECM and in what the scan tool reports. But, thanks to the law of averages, the short-term fuel trims that are reported on the scan tool are a good sampling representation of what the ECM is doing.

Finally, it is common for different readings to occur for each bank of short-term fuel trim corrections. A split or imbalance in the trims can occur from placement of the injectors in relation to the fuel supply, location of the exhaust air/fuel sensors, configuration of the exhaust system, or even the configuration of the intake manifold. Since most ECMs can control each bank simultaneously, or even individual injector trims, this is not an issue. If the splits of the short-term fuel trims are very far apart from one another (more than ±6 percent),

Some scan tools allow the mechanic to reset the historical fuel trims. Whenever faulty components that have contributed to incorrect fuel trims are repaired or replaced, the fuel trims should be reset. This allows the proper data to be used to rebuild the historical fuel trim data.

then a mechanical or sensor issue may need to be investigated.

The location of the oxygen sensor and operating temperatures are important. The oxygen sensors should be located so the exhaust stream is a balanced sampling of all of the cylinders on that bank. If the sensor is biased to one exhaust port, the resulting oxygen sensor values could report lean when that is not necessarily the case. The orientation of the oxygen sensor is also important. Typically, the oxygen sensor should never be located at the bottom of the exhaust pipe. This helps keep dirt and debris from contaminating and possibly ruining the sensor.

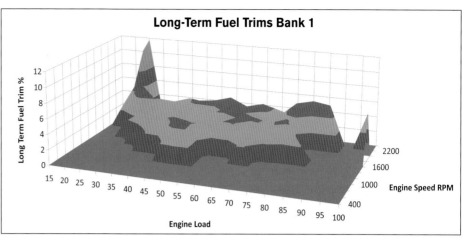

Figure 12.5. Long-term fuel trims for bank 1 are based on engine load and engine speed. The long-term fuel trims show the historical fuel correction, which can normally vary ±6 percent. It is very rare for the long-term fuel trims to be exactly 0 percent. In this graph, areas that are shown as 0 percent are combinations of engine speed and engine loads that have not been reached, because a scan tool reset the fuel trims or because the loads have never been reached by the engine.

Historical (Long-Term) Fuel-Trim Adjustments

As the ECM continues to make real-time adjustments to the injector pulse widths for the fueling, the ECM keeps track of these changes over time. The benefit of keeping track of these changes is to use the law of averages to limit the amount of real-time changes that are required.

Imagine if your family has long enjoyed a certain barbeque recipe for chicken. They always preferred it a little more spicy but not as salty. So, when mixing the ingredients for the recipe, you know that you always remove a teaspoon of salt per pound of chicken, and add 2 teaspoons of chili seasoning per pound of chicken. Instead of sampling the chicken all the time while making it and adding the chili and not adding salt, you would automatically

Flex-Fuel Vehicles

If the vehicle being diagnosed is considered a flex-fuel vehicle, it means that the manufacturer has designed the vehicle to run on two or more different fuel types. When diagnosing the vehicle, it helps to know what type of fuel the vehicle is currently using without having to draw a sample of the fuel and test it. Fortunately, the ECM can report the type of fuel that the vehicle is currently using. By using the scan tool, the PID 0x51 in Mode 01 can be queried, and it returns a single byte of data that shows the current fuel type being used. Even non-flex-fuel vehicles report their single fuel type. The following code is returned to indicate the fuel type:

0x01	Gasoline	0x07	Propane	0x0F	Bi-fuel running electricity
0x02	Methanol	0x08	Electric	0x10	Bi-fuel mixed gas/electric
0x03	Ethanol including E85 and E100	0x09	Bi-fuel running gasoline	0x11	Hybrid gasoline
		0x0A	Bi-fuel running methanol	0x12	Hybrid ethanol
0x04	Diesel	0x0B	Bi-fuel running ethanol	0x13	Hybrid diesel
0x05	Liquefied petroleum gas (LPG)	0x0C	Bi-fuel running LPG	0x14	Hybrid electric
0x06	Compressed natural gas (CNG)	0x0D	Bi-fuel running CNG	0x15	Hybrid mixed fuel
		0x0E	Bi-fuel running propane	0x16	Hybrid regenerative

DYNAMIC FUEL CORRECTION

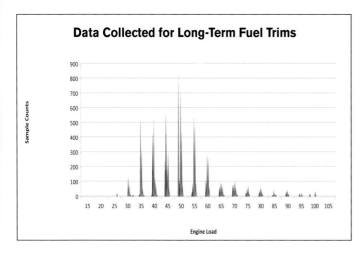

Figure 12.6. The long-term fuel trims are historically based; the more times the trims are sampled, the better the data stored in the ECM. As expected, this graph shows engine loads in the normal driving ranges (manifold pressures between 30 and 60 kPa).

adjust the recipe before starting. Thus, if you were cooking 2 pounds of chicken, then you'd add 4 teaspoons of chili, and you would not add 2 teaspoons of salt to the recipe.

The ECM controls fuel in much the same way. It keeps track of all fueling changes that occur from the short-term fuel trims. These changes then are constantly being averaged, and this average makes up the historical trim known as the long-term fuel trims. The long-term fuel trims work similarly to the short-term fuel trims in that they are expressed as a percentage of deviation from the commanded air/fuel ratio as witnessed in the exhaust stream.

Figure 12.5 shows the amounts of sampling in the graph of 12.6. The ECM is constantly saving the short-term fuel trims and adding that to the running average of the long-term fuel trims. The more data points that exist for a particular condition, the better off the long-term fuel trim average is going to be. When the sample contains thousands of data points, a single discrepancy does not have a significant effect on the overall long-term fuel trim. But when the battery is disconnected or the ECM is reset, all of the data contained in the long-term fuel trim tables may be reset, depending on the manufacturer. The tables then have to be rebuilt through drive time. Once the long-term trim tables have been reset, many mechanics advise to drive the vehicle for 50 miles or more, so the tables can be repopulated.

Repopulating the table is also very important when a sensor in the system that can affect the short-term or long-term fuel trims has gone bad. If the sensor was relaying incorrect data, then the entire table can possibly be incorrect. Many scan tools can reset the fuel trim data back to zero, and this is a very important feature to have. If the data in the long-term fuel trim table has been wrong for a long time, it may take some drive time to accumulate all the necessary data and correctly adjust the fuel trim. After the scan tool resets the long-term fuel trim tables, the engine should start running normal quickly and only have to trim a slight amount. It is far better to have a long-term fuel trim table set to zero than it is to have it set to 25-percent fuel addition!

DTCs Related to Fuel Trims

The short-term and long-term fuel trims are the essential features in the feedback loop for the fueling system. Both fuel trims rely on a multitude of sensors, so the system can provide accurate fueling corrections. Thankfully, the OBD-II system also provides for a multitude of checks and balances, which include the sensors that the fuel trims rely on to give accurate readings. The OBD-II system also diagnoses when the information regarding the fuel trims can be faulty. The following are faults that can appear when generated by errors in the fuel trim system:

- P0170   Fuel Trim System Bank 1
- P0171   Fuel Trim System Lean Bank 1

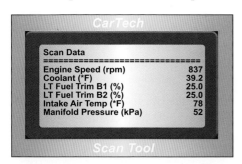

A scan tool shows both banks as reading lean after setting a P0171 and P0174 DTC.

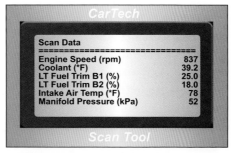

A scan tool shows both banks as reading lean. Bank 1 has set a P0171 DTC. And although Bank 2 is reading very lean, it has not surpassed the threshold to set the P0174 DTC yet.

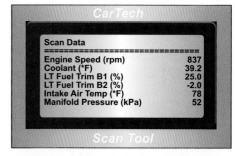

A scan tool shows bank 1 as reading lean after setting a P0171. Bank 2 is shown within specifications being 2-percent rich.

CHAPTER 12

- P0172 Fuel Trim System Rich Bank 1
- P0173 Fuel Trim System Bank 2
- P0174 Fuel Trim System Lean Bank 2
- P0175 Fuel Trim System Rich Bank 2

P0171 and P0174 DTC Fuel-Trim System Lean

A hard limit determines how much the long-term fuel trims can adjust fuel to compensate for a lean condition. Once this limit is reached, the DTC is set and the MIL is illuminated. If the DTC is being set because the maximum amount of fuel is being added, the diagnoses needs to include checking the sensors that dictate airflow and exhaust air/fuel ratio monitoring, as well as check the health of the fuel system itself.

The short-term and long-term fuel trims should be checked with a scan tool. If only one bank is shown as lean, then the incoming airflow sensors and calculations are most likely functioning properly because an error here affects both banks and not a single bank. If both banks are showing lean, or if one bank has set the DTC and the other bank is showing very lean but not enough to set the DTC, then the exhaust air/fuel ratio sensors can typically be ruled out because it is unlikely that both sides would go out simultaneously.

When you encounter either (or both) P0171 or P0174 DTC, use the scan tool to analyze the incoming airflow. If the vehicle is equipped with a MAF meter, which measures the incoming airflow, you can use the scan tool to analyze the reported incoming airflow. A MAF reports a frequency or a voltage, but the ECM converts this data to an airflow quantity. The MAF's values reported for airflow are often low or nonexistent. If the MAF has failed, DTCs are set to report that fact. The MAF often reports an airflow that's lower than the actual airflow coming into the engine. So the fuel calculations are often low, making both banks lean, and the DTC is set.

If the MAF readings appear accurate on the scan tool, a quick test to verify the readings is to look for unmetered air getting into the intake. Remember that the ECM is counting on an accurate measurement of incoming air going through the MAF. If there is a leak after the MAF, and the engine is drawing in unmetered air, the ECM does not measure this and, as a result, it does not add enough fuel for the additional unmetered airflow. A quick squirt of brake fluid or propane gas alters the idle of the engine and helps pinpoint the leak.

Another common cause for these two DTCs is failure of the exhaust air/fuel ratio sensors. The oxygen sensors are designed to fail lean. So when the sensor fails, it reports that the air/fuel ratio in the exhaust requires more fuel. The ECM then continues to add fuel until the DTC is set. A quick scan of the activity on this sensor shows that the sensor is not functioning properly and should be replaced. Typically, when one bank has failed lean and the other bank is reporting normal operating conditions, the exhaust sensor on that bank has failed. An exhaust leak prior to the location of the exhaust sensor can also cause the exhaust sensor to report excess oxygen, and thus report a false lean problem. The exhaust system should be inspected carefully.

The most common cause of the two DTCs is a physical problem with the fuel delivery system. The ECM has calculated the fuel delivery capabilities of the fuel injectors based on a preset fuel pressure coming from the fuel pump. When there is a blockage or restriction, the fuel injectors do not see the proper fuel pressure, and as a result, flow less than the ECM has prescribed. Therefore, when the ECM calculates the amount it needs—for example, a fuel injector pulse width of 4.27 milliseconds to deliver the properly metered fuel—the actual fuel that flows is much less than needed. It doesn't take long for the fuel trims to swing to the lean DTC codes under these circumstances. A clue that this is an issue is when the incoming airflow still checks out, and because the exhaust air/fuel sensors are operating normally. A quick test with a fuel-pressure gauge shows if the pressure is being delivered at the factory requirements. A clogged fuel filter or a bad fuel pump may be the cause of inadequate pressure. In some cases, multiple injectors could be bad, but this typically shows up as a misfire.

P0172 and P0175 DTC Fuel-Trim System Rich

A rich condition is the other, but much less common, side of fuel-trim error codes. Just like the lean code problems, the airflow is the first thing to look for with a scan tool. The mechanic looks at the MAF's reported airflow to see if it is reporting properly. If the vehicle is not equipped with a MAF, then it is a speed-density-equipped vehicle, which uses the manifold pressure sensor, intake air temperature sensor, and throttle position sensor to determine the airflow. Check all of these sensors to ensure that they are reporting in the proper range.

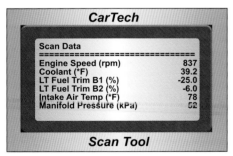

A scan tool showing bank 1 as reading rich after setting a P0172. Bank 2 is shown as being within specifications and is shown as 6-percent rich.

CHAPTER 13

ENGINE IGNITION CONTROLS

Ignition is part of one of the four cycles of the Otto engine model. In the combustion cycle, the spark plug ignites the volatile air/fuel charge in the cylinder and work is done. This ignition event is timed according to the piston's location relative to top dead center (TDC). Thus, a relationship is established to mechanically or electrically determine where the piston is positioned when that ignition event is triggered to initiate the combustion cycle. Several key components make up the ignition system and, as technology has progressed, many of the older components have been replaced with new ones having better technology.

What is Ignition Timing?

Ignition timing essentially defines the point at which the spark plug is fired in relationship to the crankshaft position during the combustion cycle of the Otto engine model. Ignition timing is typically expressed in terms of the angle in degrees based on the TDC or bottom dead center (BDC) of the piston position and/or crankshaft position.

The concept behind ignition timing is the control of the flame front and combustion burn. After the intake charge has been drawn into the combustion chamber and compressed, a very volatile mixture of air and fuel is ready for combustion. The ignition cycle introduces a spark to this charge via the spark plug. The tiny spark that jumps across the gap of the spark plug starts a multitude of flames spreading in different directions as the volatile air/fuel charge burns. Many people mistakenly think that the ignition burn is akin to a large ball of fire within the combustion chamber. In actuality, tiny fingers of flame advance outward in all directions, starting from the spark plug or ignition source, and fan out across the combustion chamber.

Engine designers commonly refer to this pattern of this combustion as the flame front of the ignition process. As the flame front advances, heat and pressure inside the combustion chamber continue to rise, typically peaking a few degrees after the piston passes TDC. After all of the air/fuel charge has been consumed and the piston travels back down the bore, the pressures resulting from the ignition exert extra force on the piston in a downward trajectory.

The key to optimizing timing is to control the relationship between peak pressures in the combustion chamber and the piston's position relative to TDC. If the peak pressures occur while the piston is still rising in the bore, the energy is wasted fighting the momentum of the piston coming up. If the ignition occurs too late, then the piston is traveling back down the bore, expanding the combustion chamber's volume, and thus bleeding off the pressure and reducing the force exerted on the piston. Factor into all of this the effect that timing has on emissions, and it is clear that timing is much more than randomly starting the combustion cycle!

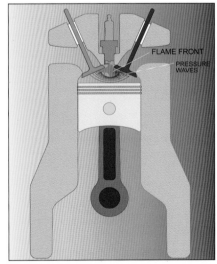

During the combustion process, the air/fuel charge is ignited and a flame front advances from the ignition point in a variety of directions. The pathways of the ignition resemble fingers of flame, not a circular-ball shape. A high-pressure wave runs ahead of the flame front, increasing cylinder pressure dramatically.

CHAPTER 13

Generating High Voltage

The modern automobile's charging system typically operates at around 12 volts, although some heavy-duty vehicles may be equipped with multiple batteries generating 24 volts or more. Unfortunately, it takes significantly more voltage to jump the gap of a spark plug. To create a spark across a distance of .050 inch (a somewhat-typical spark plug gap) can take 10,000 volts or more. So the ignition

Various Spark Systems

The most basic spark system has a single coil that is shared among all of the cylinders and spark plugs. But, as technology has evolved, a variety of different styles of ignition have emerged. Here are a few of the more popular styles:

Coil on Plug

This is a modern adaptation of the single-coil approach, where each cylinder has its own dedicated coil to fire its spark plug. On many of these implementations, the coil plugs directly onto the spark plug, or is connected with an extremely short spark-plug wire. The ECM can control the ignition fire time of each cylinder individually, which allows the ultimate tuning capability for emissions and performance.

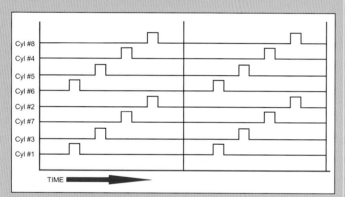

This representation of a wasted spark system shows how multiple spark plugs are fired on an engine using wasted spark ignition. On this engine, the firing order is 1-3-7-2-6-5-4-8. When cylinder number-1 is fired, cylinder number-6 is also fired, but it is in a cycle state such that the spark does not trigger combustion in that cylinder. The next cylinder to ignite is cylinder number-3. At the same time, the coil fires voltage to cylinder number-5, which has no affect on it. As the firing order continues, cylinder number-7 comes upon the combustion cycle, and the coil fires voltage to its plug. At the same time, cylinder number-4 also gets a spark, but because it is not on the combustion cycle, it is not affected.

Multiple Spark Discharge

Initially patented under United States Patent number-3926165, issued in 1975, the multiple-spark-discharge system takes advantage of the fact that under a single spark system, the entire air/fuel charge is not completely burned during the ignition process. To optimize and completely burn the air/fuel charge, the multiple-spark-discharge system fires spark into the cylinder multiple times during the combustion process. By firing the spark plug multiple times, the unburned air/fuel charge is more efficiently and completely burned, which improves emissions quality and performance.

Wasted Spark System

The term wasted spark accurately describes this style of ignition system. Typically, the ignition system has one coil for every two cylinders. As a result, multiple cylinders share the same coil. The single coil fires both spark plugs simultaneously. The design of the ignition system is important because one cylinder will be in its combustion cycle while the other cylinder that shares the coil will be in the exhaust cycle. This means one of the spark plugs is fired at a time that the spark is wasted; the spark plug does not initiate a combustion cycle. The main advantage of a wasted spark system is that it performs better than a single coil system, and it is much less expensive than a true coil-on-plug system.

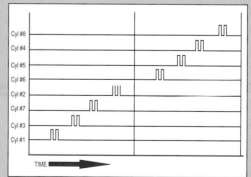

This representation of a multiple spark discharge system shows how the coil has multiple spark occurrences on the same combustion cycle.

ENGINE IGNITION CONTROLS

system has to significantly increase the amount of voltage sent to the spark plug in order to create the spark to initiate the combustion process.

Electrically, when voltage levels must be changed, a component called a transformer is used. When voltage needs to be decreased, a step-down transformer is used. And when voltage needs to be increased, a step-up transformer is used. A transformer uses electromagnetic induction, in which a coil of wires and a magnet create voltage as the magnetic field is changed or moved. Introducing a current in a coil of wires generates this magnetic field.

In the automotive ignition application, the method employed is a step-up-style transformer that uses a pulse model to generate a significant voltage step-up. Inside the automotive pulse transformer (commonly called the coil) are two coils, one inside the other. The outer (primary) coil consists of a few hundred windings, while the inner (secondary) coil consists of several thousand windings. The primary coil applies a current to charge the coil. As the current is applied, a magnetic field grows around the primary coil. At a determined time, the current flowing into the primary is abruptly halted, which causes the magnetic field around the primary coil to suddenly collapse. This sudden magnetic field change, which also encompasses the secondary coil, causes a surge in current in the secondary coil. This causes a sudden high-voltage spike (40,000 volts or more) coming out of the secondary coil.

With the sudden collapse of the magnetic field, the primary side of the coil is affected. A sudden surge of voltage flows backward from the primary coil to the automotive system. Fortunately, this isn't as much voltage as seen on the secondary coil, and a capacitor controls and stores the excess voltage, which can slowly be redirected back into the system. On older automotive systems, this capacitor (also known as the condenser) was located inside the distributor.

Distributors

Before computer controls were created, a distributor and a coil were the primary system that controlled the engine's ignition. The distributor is aptly named for its function in the ignition system. Its primary job is to distribute the ignition voltage to the cylinders in an organized and timed manner. The trigger that

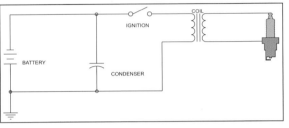

This schematic representation of how the high-voltage spark is generated at the plug shows the two windings of the coil. The primary side of the coil is tied to the 12-volt system, while the secondary side of the coil is tied to the spark plug. When the ignition switch (or points in older systems) is turned on, current flows through the primary coil. This charges it and creates the magnetic field around the primary and secondary windings. When the switch is opened, the current flow is halted, and the primary field collapses. A sudden high-voltage spike on the secondary side of the coil is the result. This brief high-voltage spike causes the spark to jump across the gap of the spark plug and ignites the air/fuel charge in the chamber.

Working with High Voltage

At some point, everyone accidently gets shocked. Working around 12-volt-DC car voltage has its dangers, but with the high voltage that automotive coils produce, it can be very painful and even life threatening in certain cases. Remember that electricity always seeks the easiest path to ground and, in most cases, the human body makes an excellent conductor. The average coil produces more than 40,000 volts to allow the spark to jump the gap on the spark plug. This high voltage also allows a spark to jump the gap between the coil and the mechanic. Take extreme care when working on ignition systems. Keep grounding paths in mind and make sure that you are not in the chain.

Never touch a coil while the engine is running. A coil that is leaking voltage can jump a spark to a screwdriver, a wrench, or even a hand. When working around a coil, use insulated gloves, such as those worn by electricians or utility linesmen. This extra insulating layer helps isolate you from the path to ground. If you think a coil is faulty, bring it to an automotive parts store that has the proper testing equipment to determine if the coil is bad. Most home enthusiasts do not have the proper equipment to detect the quick discharge of high voltage from a coil, nor do they have a multimeter capable of measuring that voltage level.

collapses the coil is an electro-magnetic switch (i.e., points) or a completely electrically controlled switch. A simple camshaft on the distributor is responsible for timing when the coil's field is collapsed. The distributor is also responsible for ensuring that the voltage is sent to the proper cylinder's spark plug. The rotor in the distributor spins and aligns itself to a pin attached to the spark plug wire through the distributor's cap.

The distributor is connected directly to the camshaft via a gear. This gear keeps the timing of the ignition system in sync with the valvetrain and crankshaft. The position of the distributor determines the timing of the ignition system. If different timing is required, the distributor itself can be rotated clockwise or counterclockwise to advance or retard timing. An inductive pickup triggers a timing light (a strobe), which detects the voltage in the spark plug wire. The timing light measures timing in degrees before or after TDC on cylinder number-1.

Again, the amount of timing is dependent on engine speed and load. The distributor employs a mechanical method or vacuum-based method (or a combination of both) to advance or retard timing. A mechanical-based system uses weights and centrifugal force to add timing based on the rotation speed of the distributor's shaft, which is connected directly to the engine's camshaft. A vacuum-based system uses engine vacuum to advance or retard engine timing. Either system gives some control over engine timing, but neither is sophisticated enough to adapt to a variety of engine conditions. That is why control of ignition timing was gradually built into the ECM, and distributors were eventually removed from most engine control systems.

Knowing When to Fire

The advent of tighter emissions standards, increased fuel mileage, and the demand for better performance meant that the distributor ignition system was not capable of performing with the precision required. A normal distributor basically knows nothing about crankshaft position or where the distributor currently is in respect to each cylinder and its place on the four-stroke cycle. Thus, the ECM can't monitor the stroke of the engine's cycle or have any control over it.

To accurately control timing, the ECM has to know which cylinder is currently on the combustion cycle. But it also has to know exactly where the crankshaft is rotationally, down to 1-degree-or-less resolution. Thankfully, sensor technology has improved dramatically over the years, as have precision machined, mass-

Optical Distributors

As an interim step between mechanical distributors and completely distributorless ignition systems, some manufacturers adopted an optical distributor. It still employs many of the features of a standard mechanical distributor, but it has much higher positional accuracy. It still uses a single, common coil and a cap-and-rotor system to distribute the ignition voltage to the proper cylinders. The cap-and rotor use a stub shaft or gear system to connect to the camshaft, so that the engine speed synchronizes with the optical distributor. The major difference is that, unlike a mechanical distributor, which is purely an output device for ignition voltage, the optical distributor reports the rotation angle of the crankshaft.

The crankshaft's position is typically tracked via a shutter wheel and a laser or light-sensitive pickup. The shutter wheel features small holes located at predefined increments. The light source is located behind the wheel. As the holes pass in front of the light, the light shines through, and a sensor on the opposite side of the wheel detects the light. If these holes are located at 1-degree increments, then the sensor is capable of detecting the position of the crankshaft within 1 degree of resolution. Most of the time, a second light source is also used, along with a different-shaped hole or slot located at TDC of each cylinder. This allows the ECM to not only be able to track the position of the crankshaft relative to angle, but to also track exactly which cylinder is at TDC.

However, because it still employs a single-coil cap-and-rotor-style distribution system for the ignition spark, the drawbacks of this style system have all but eliminated it from modern vehicles. Most important, the shortcomings of the optical pickup make this system a lot less reliable than its current-day counterparts. The holes that are used to indicate position are extremely small and prone to fouling and clogging. If a few of them get fouled, the ECM's counts are off; or worse, the entire firing order is incorrect. Also, because the optical-distributor ignition system contains a rotor-and-cap, the entire unit must be vented of the ozone created by the arcing of the rotor to the cap. This ventilation system can cause more contaminants to enter into the distributor and foul the optical sensors.

ENGINE IGNITION CONTROLS

produced parts. The combination of these two technologies has allowed automobile manufacturers to design the ECM to constantly monitor the position of the crankshaft at all times. This is the key for improved timing and misfire detection.

Most automobile manufacturers employ a sensor and a reluctor wheel mounted to the crankshaft. This wheel is precision made with slots or notches

The crankshaft reluctor wheel for the General Motors LS-1 engine is accurately pressed onto the crankshaft. The combination of the two wheels together and their various staggered teeth lets the ECM know exactly the rotation angle of the crankshaft as well as the cylinder that the engine is on.

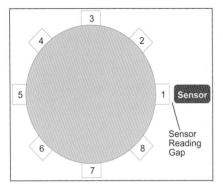

This crank reluctor wheel is a simplified version of the actual part. In this case, there are eight individual tabs on the wheel that represent the eight cylinders in this engine. The sensor can detect the tabs because they are close enough to alter the magnetic field of the sensor. As the wheel rotates the tab away from the range of the sensor, the magnetic field changes and resumes its normal values, so the sensor does not register the tab until the next cylinder tab comes by. Typically, most reluctor wheels have an indexing system. It identifies a particular cylinder, so the ECM knows where the crank position is at all times.

Hall Effect Sensors

A common method for measuring the position of an object is to take advantage of a phenomenon known as the Hall Effect. The principle behind the effect is: When a current is passed through a conductor placed in a magnetic field, the magnetic field exhibits a force on the charge carriers and forces them to one side of the conductor. This pushing of the charge carriers to one side allows a small amount of voltage to be measured. The stronger the magnetic field, the larger the voltage can be. These voltages are very small, and typically need to be amplified to make them useful. The Hall Effect sensor does just this. It can measure that small voltage and then amplify it so the ECM, or any other controller for that matter, can detect the presence or position of an object. In most automotive applications, the Hall Effect sensor is used much like a switch; it is either on or off. So if it detects the object in question, it turns on; otherwise, it turns off.

The benefit of using a Hall Effect sensor is that it isimmune to contamination from dirt, oil, coolant, and most other things found in and on an automobile. They also are very small, allowing the manufacturer to install these sensors in very tight confines. With no moving parts, there is never an issue with them wearing out over time or losing accuracy. The downside of these sensors is that they can be influenced by adjacent magnetic fields. Wires or other inductors can generate magnetic fields and can interfere with the sensor's magnetic field. Furthermore, magnetically charged contaminants may adhere near the sensor, which can also affect its accuracy. These typically can be cleaned off and the sensor reinstalled.

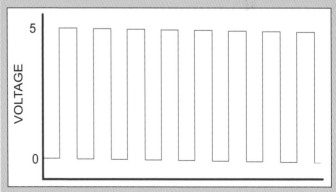

The Hall Effect sensor returns a digital logic "1" or a digital logic "0," depending on whether or not the object to be sensed is within range or not. In the automotive application, the Hall Effect sensor uses 5 volts to represent a detection state. As the crankshaft rotates, the sensor detects the reluctor slots and toggles the state of the sensor's output. The frequency of the pulses vary with engine RPM.

CHAPTER 13

machined into the wheel at precise locations. One or more stationary sensors are mounted near this wheel and can detect the presence of the slots/notches on the reluctor wheel. There is a variety of coding methods for the location of the notches on the reluctor wheel, but the basic idea behind them is that they allow the sensors to inform the ECM of the exact crankshaft and camshaft position. Furthermore, two different styles of sensors can be employed at the discretion of the automobile manufacturer: Hall Effect sensors and magnetic sensors. Both deliver crankshaft positional data but use two different detection methods.

Most vehicles require a crank-position sensor, but they also require camshaft position sensors because the crankshaft can't differentiate what is going on with the valvetrain. Therefore, the crank position does not completely reflect which cycle the cylinders are on. And remember that the crankshaft makes two complete revolutions per the camshaft's single revolution. So the ECM also uses a magnetic proximity or Hall Effect sensor connected to the camshaft, typically on the drive gear, to track the position of the camshaft. This permits the ECM to calculate exactly where the crankshaft is in relation to the cylinder cycles.

Controlling the Timing

Unlike a distributor's limited control of the timing curves of an engine, the ECM has complete control of the timing of an engine based on a variety of factors. This allows the ECM to have multiple timing tables based on fuel grades, coolant temperature, intake air temperature, engine load, engine speed, idling, etc. This helps to boost engine performance and minimize emissions.

The base timing map is developed at running conditions (non-closed throttle) based on engine speed and the amount of air coming into the engine. The timing varies a significant amount over the engine's operating range, from a high of 42 degrees of timing to a low of 7 degrees of timing. OEMs developed this table to optimize cylinder pressures, emissions, and reliability. Although, it may not result in the absolute pinnacle of performance that can be achieved with a custom ECM calibration, it does a very safe timing range for the various gasoline grades that customers may use. It also factors-in the maintenance, or lack thereof, that some engines might be subjected to. After all, the OEMs like to have engines run well past their 100,000-mile odometer mark.

Sensors that Affect Timing

The timing map uses engine speed and airflow calculations, so the sensors must provide accurate information. If one of these sensors gives inaccurate information, then incorrect timing will

Magnetic Proximity Sensors

A magnetic proximity sensor registers the change in a magnetic field when a ferrous object is introduced into the field. As the sensor "sees" the object approaching, it increases voltage toward its peak voltage (in the automotive application, this is +5 volts). As the object goes away, it reduces the voltage by -5 volts before returning to 0 volts. Magnetic proximity sensors do not require magnets to be mounted to the object being measured, unlike the requirement of a Hall Effect sensor. For this reason alone, it is much easier to implement precision measurements with the magnetic proximity sensor versus the Hall Effect sensor. And like the Hall Effect sensor, the magnetic proximity sensor is impervious to dirt, oil, coolant, and most contaminants found in/on an automobile.

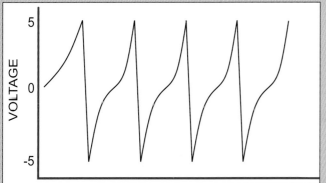

The magnetic proximity sensor returns an alternating voltage based on the sensing of the object in question. As the crankshaft rotates, the sensor returns a signal that alternates toward 5 volts as it comes into range, and down toward -5 volts as it moves away. The amplitude and frequency of the signal vary with engine speed. Magnetic proximity sensors can be tested to verify if they are working by measuring the resistance between the two terminals. If infinite resistance (an open circuit) or a short (0 ohms) is measured, then the sensor is typically deemed to be faulty.

be derived. In a best case scenario, performance and emissions suffer, and in a worst-case scenario, engine damage may occur. Most scan tools report the amount of timing being commanded by the ECM. If timing values look out of spec, use the scan tool to quickly check sensor values to make sure that they are delivering proper data. If the engine speed is showing incorrect data, check the crankshaft position sensor (CKP) and camshaft position sensor (CMP). If the airflow calculations are suspect, check the MAF (if equipped) as well as the MAP sensor for correct readings.

Intake air temperature also plays a part in timing calculations. The temperature of the incoming air always affects the density of the air charge in the cylinder. So timing must be adjusted according to that incoming air temperature. The ECM samples the incoming air charge via the IAT sensor. Based on the engine's airflow calculation and the intake air temperature's reporting, the ECM decides if it needs to remove timing in order to protect the engine from detonation, which could cause permanent damage. If timing is shown to be lower than expected, check the IAT sensor for accurate readings.

The internal temperature of the cylinder chamber is important for emissions as well as performance. When a cold engine is first started, the fuel is richened in order to give a better burn. This requires that timing be added to the base timing tables. The ECM scans the ECT sensor to determine the temperature of the coolant in the engine, and thus the rough temperature of the engine itself. By referencing a lookup table in the ECM, the timing is adjusted to compensate for the engine temperature. If the timing shown on the scan tool is more than expected, the data coming from the ECT sensor should be verified. Since this particular data adds timing to the base timing tables, damage can occur by running too much timing in a non-cold-engine scenario.

An engine at idle typically is under very little load; the drivetrain and accessories require power along with the power required to turn the reciprocating parts of the engine. The engine can then operate on a lot less timing than under load. This keeps internal cylinder temperatures down and emissions in check. The ECM typically references two sensors to determine if the engine is in an idle state. First, it checks the TPS, which indicates the amount the throttle is open. If the throttle is at the spec'd idle position, then it also checks the speed of the engine. Based on these readings, along with the current airflow demands of the engine, a new timing map is used in place of the standard timing maps.

Some ECMs differentiate whether the transmission is in gear or in park and employ different timing tables because of the increased loading from the drivetrain and power requirements. Typically, these two tables do not differ by much. If timing shown on a scan tool is different than expected, and the engine seems to idle higher, check the TPS to make sure that the ECM is commanding the different timing map for closed-throttle timing.

Detonation and Pre-Ignition

Detonation, pre-ignition, spark rattles, and spark knock are all terms for out-of-control cylinder burns, and none are good for the health of the engine. The key to eliminating these problems is to understand what causes the reaction in the chamber. The two different types of out-of-control cylinder burns are detonation and pre-ignition. Each has a different root cause, but the outcome of both can lead to broken piston ring lands, failed spark plugs, blown head gaskets, damaged bearings, leaking valve seats, and eventually overall engine failure. Luckily, there are safeguards built into the ECM to help curb some of this, but they can only go so far.

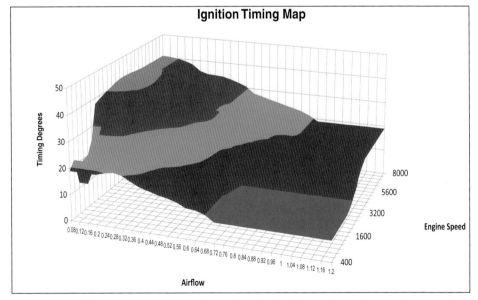

The ignition timing map is based on the calculated airflow and engine speed in RPM, and timing varies to optimize performance, reliability, and emissions. The base timing map can be altered by input from other sensors, which is based on the operating conditions the engine is recognizing. Continual detonation may also affect this timing map, and it can result in a significant reduction in values for safety.

ECM Evolution

To understand the impact that OBD-II, emissions requirements, fuel-mileage requirements, and computer performance has had in the evolution of the ECM, you need look no further than the timing tables. A brief comparison between a timing table for a 1986 Ford Mustang and a 2010 Chevrolet Corvette is striking. Due to the limited amount of memory and the operating-speed constraints in the processor, the Ford ECM has a total of 80 timing settings. Contrast that to the Chevrolet ECM's total of 1,089 different timing settings.

The difference is the amount of resolution that the newer ECM can offer. It can handle small, precise adjustments in timing over a vast range of engine speeds and engine loads. The older ECM never had the capability to cover that many different operating conditions and as a result, it had to take a best-average approach to developing the timing maps. As a result, the older ECM cannot compete with the newer ECM in terms of fuel economy and emissions control.

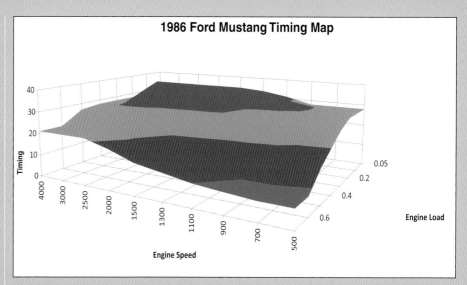

The ECM controls a limited amount of engine points on the ignition timing map for a 1986 Ford Mustang. The limited number of data points means that the ECM has to extrapolate in operational areas that fall in between the points in the table. This may lead to less fuel economy and increased emissions.

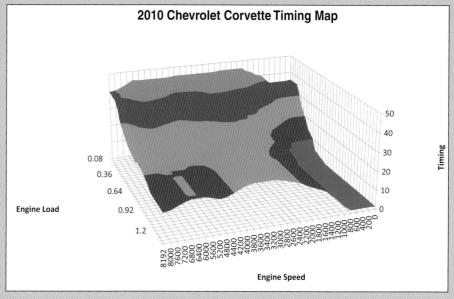

The ignition timing map for a 2010 Chevrolet Corvette contains enough resolution to allow the ECM to adapt to just about any operational parameter the vehicle may encounter. By optimizing each point for performance, fuel economy, and emissions, and by having a very small difference between one operating parameter to the next, the movement from one table entry to the other is very smooth and causes little in terms of detrimental operating conditions.

ENGINE IGNITION CONTROLS

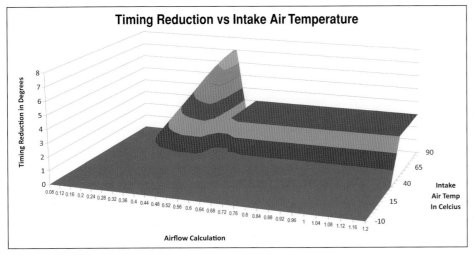

If the IAT sensor detects that the incoming air temperature is rising, timing is reduced from the base timing map to protect the engine from detonation.

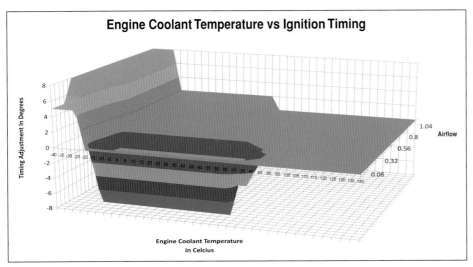

A colder engine temperature requires more fuel and more timing. As the engine heat rises, the timing is slowly withdrawn. This table shows the amount of timing, based on engine coolant temperature, that this specific ECM adds to the base timing maps.

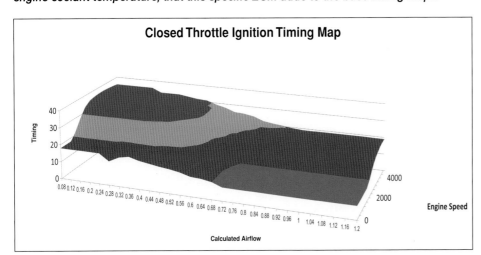

The illustration on page 85 shows normal flame-front advancement in the chamber. A single flame kernel is generated from the ignition source (spark plug) that spreads throughout the combustion chamber. It also shows the pressure wave and precedes the flame front fanning out in the combustion chamber. A graph of temperature and pressure in the chamber would show both of them smoothly increasing to peak pressures.

Another aspect of the ignition process is the gas boundary layer that forms at the instant of flame advance. This layer, only several molecules thick, is much like a coating that exists in the combustion chamber. During the ignition process, the combustion chamber temperatures reach almost 2,000 degrees F, which is well above the melting point of aluminum (about 1,220 degrees F). The boundary layer provides thermal buffering so that the aluminum engine parts don't melt or get damaged.

Detonation or pre-ignition indicates a loss of flame-front control. It means that additional, but smaller, flame kernels are occurring outside the ignition-induced flame kernel. These smaller flame kernels have their own pressure waves, which emanate radially.

When these pressure waves from the multiple flame fronts collide, they cause a detonation, which makes an audible

If the throttle is in a closed position, a new timing table is used, rather than the base timing table. The engine speed covers the spectrum of values because the throttle is closed during this driving situation. For example, when decelerating, the throttle can be closed, but the engine speed can be above idle speeds. By adjusting the timing, cylinder pressures can be manipulated and, more importantly, engine emissions can be controlled to ensure a complete air/fuel burn during these situations.

noise. This wave is similar to a ping from a submarine sonar system colliding with another object, and "pinging" aptly describes this phenomenon. Furthermore, as these waves collide and cylinder pressures spike, the gas boundary layer breaks down and leaves the mechanicals unprotected and prone to damage. The sudden excess heat has been known to melt completely through the top of a piston, damage piston ring packages, and cause heads to lift from the engine block.

The two types of pinging are pre-ignition and detonation. Detonation occurs when excessive heat and pressure cause the air/fuel charge to abnormally auto-ignite, and then sub-flame kernels are generated. Pre-ignition can occur when a hot spot within the cylinder causes the air/fuel charge to ignite prior to spark plug ignition or when there exists a very lean air/fuel ratio in the chamber. Pre-ignition is typically a mechanical issue while detonation is typically an ECM-controlled or driving-condition issue. Pre-ignition is the more dangerous of the two because pre-ignition can occur while the piston is still traveling up the bore. This causes an opposite force pushing down on the rising piston, which can damage bearings, crankshafts, pistons, and connecting rods. Although automotive OEMs find a light amount of detonation acceptable, heavy detonation eventually causes catastrophic engine damage and failure, and no engine is ever designed to accept any amount of pre-ignition.

The ECM and Detonation

The ECM controls the ignition timing and, indirectly, it also controls what occurs when an engine experiences detonation. Again, detonation is actually pressure waves colliding together in the combustion chamber. The wave resonates throughout the engine block, much like a sound wave that passes through the walls of a house. So, early on, the ECM needed a sensor to be able to detect this wave and discern if it is truly detonation or some other mechanical noise.

Midway through the OBD-I ECMs (about 1989), the knock sensor was introduced. It is basically a piezoelectric crystal sensor, or piezo sensor for short, which acts similar to a microphone, listening for frequencies emanating from the engine block. The knock sensor constantly sends reports to the ECM based on the frequencies that it is recording. The ECM studies this signal and determines if the recorded frequency represents true engine detonation waves, some other frequency generated from the rotating parts of the engine, or external frequencies, such as road noise. The ECM can reference its software tables to help eliminate false positives for detonation. The ECM always errs on the safe side and can initially react to these false positives and check for ongoing correction. It does not have the intelligence to learn from its mistakes though, so if the frequency continues to be picked up at intervals, it continues to treat this false positive as detonation.

When the ECM detects and determines that the response from the knock sensors is an actual detonation event, it first tries to remove timing. The ECM continues to remove timing until the knock sensors start to show reduced knock occurring. Because the pressure waves resonate throughout the engine block, the waves slowly dissipate over time, much like the water ripples from throwing a rock in a still pond. The ECM constantly monitors the knock sensor and slowly adds timing from its timing table. If the knock sensors start to pick up additional detonation activity, the ECM again reduces the commanded ignition timing to eliminate the detonation.

If the detonation continues to occur as timing is added, some ECMs permanently adjust the timing at those points to avoid any more detonation the next time the engine enters that operating condition. This reduction in timing may gradually be removed over an extended time period, or when the vehicle is refilled with gasoline. The premise behind the refill is the fact that the vehicle may have just a tank of bad gasoline, or, in the case where the engine requires a higher-octane gasoline, the tank was filled with a lesser grade of gasoline.

What Causes Detonation and Pre-Ignition?

Some of the more common causes of detonation and pre-ignition include:

- Some vehicles are manufactured to use higher-octane-rated fuels. The higher octane helps to reduce

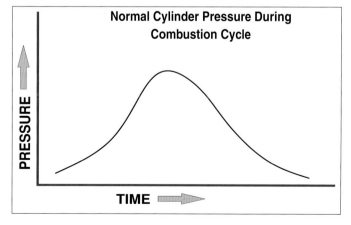

This is a graph of the normal pressures seen during a combustion cycle. It is very smooth as it tracks to the peak cylinder pressure and slowly comes back down to finish the cycle.

ENGINE IGNITION CONTROLS

the tendency to detonate in these engines. If a lower octane fuel is used, the engine is more prone to detonation. The manufacturer's recommended fuel grades should always be followed.

- A defective knock sensor can prohibit the ECM from reducing timing, making it more audible to the operator. The ECM monitors the knock sensor to ensure that it is giving valid data. A damaged knock sensor should set one of several DTCs and illuminate the MIL on the dashboard.
- The EGR system may be defective. The valve itself may be stuck or inoperable, or one of the hoses may be leaking.
- The engine's operating temperature is too high. Higher cylinder temperatures equate to higher cylinder pressures and are more prone to detonation.
- High intake-air temperatures create hotter intake-air charges and are prone to detonation. Ensure that the intake hoses/tracks are not allowing hot air to mix with the incoming air charge. Also, a vehicle that has been sitting and running may have retained extra heat in the intake and may be prone to detonation until the air charge can cool.
- Improper fuel delivery causing a lean air/fuel condition may make the engine prone to detonation. The long-term fuel trims show the lean condition and may have already set a DTC. Plugged fuel filters or bad fuel pumps are the leading cause of this condition.
- High-mileage engines that have years of carbon buildup in the combustion chamber can cause hot spots and pre-ignition conditions. Some top-end engine cleaners may help alleviate this problem, although there is no "miracle in a can" that is an absolute cure.

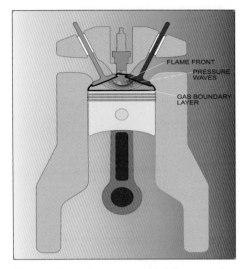

The gas boundary layer, which is several molecules thick, forms within the combustion chamber to protect the engine components from damage. The boundary layer is shown on this drawing as the black coating in the chamber.

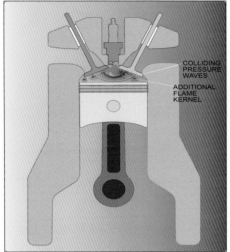

During detonation, multiple flame kernels can occur at various points in the combustion chamber. Each of these kernels produces the pressure wave emanating from the flame kernel. The collision of these waves creates the spike in pressures. The multiple pressures also break down the gas boundary layers, which leave the pistons and cylinder surfaces unprotected from the increase in temperature.

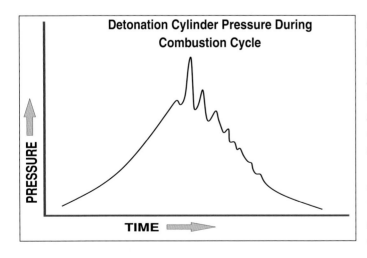

During detonation in a combustion cycle, the cylinder pressures spike. In this graph, cylinder pressures suddenly spike upward when detonation starts. This spike of pressure greatly exceeds the normal cylinder pressures seen without detonation.

When detonation or pre-ignition gets too severe, physical damage can occur to the piston, which can cause a complete engine failure. This photo shows the path of the hot combustion flames where they have burned through the top of the piston and ring land package. (Photo courtesy Jim Daley Westech Automotive)

Detonation and its Effect on Power

Detonation is harmful to the life of the engine, but what about its effect on power and torque?

The graph shows the dynamometer readings of an engine that makes a clean run and a run where detonation occurs. The detonation started to occur at approximately 5,250 rpm and never diminished. As soon as the detonation occurred, power and torque rapidly dropped off. The air/fuel ratio dropped to a very rich condition because of the misfiring occurring in the cylinder. The extra fuel added by the ECM at WOT was no longer being consumed by the flame front. The ECM detected the amount of detonation on the order of 6 degrees of knock retard and it was not audible by human hearing. The power loss was on the order of 30 to 40 hp, which is definitely detectable to someone while operating the vehicle.

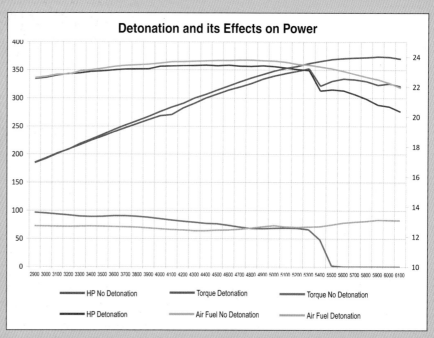

This graph shows a dyno test of an engine. One test is a clean run and detonation was present in the other run. When detonation occurred, the engine immediately lost horsepower and torque.

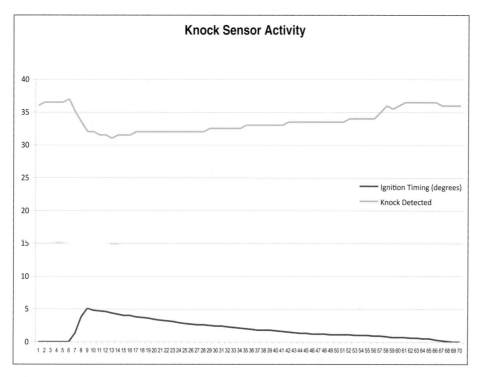

The knock sensor picks up activity during this engine monitoring session. The activity prompts the ECM to immediately retard the commanded ignition timing in order to see a reduction in knock sensor activity. When the activity starts to decay, the ECM adds back in the timing unless more activity is detected.

CHAPTER 14

MISFIRES

Engine misfires can be some of the easiest or some of the most difficult problems to diagnose. A misfire can be as drastic as a cylinder not firing at all, or it can be as subtle as an occasional misfire every couple of seconds, or even a cylinder not operating at 100 percent of capacity. The ECM has little to do with correcting a misfire issue, but thanks to OBD-II, the ECM can really aid in diagnosing and correcting the issue.

Misfire Types

There are basically two different types of misfires: the constant misfire and the random misfire. The constant misfire is easier to diagnose. With this type of misfire, the cylinder typically makes no power at all, which is also known as a "dead cylinder." In this situation, the misfiring cylinder fails to ignite the air/fuel charge, or does not transfer the energy released during the combustion into mechanical work through the crankshaft. Typical symptoms of a constant misfire are a lack of power, rough idle, increased emissions, and low fuel mileage.

The random misfire is more difficult to diagnose. This type of misfire occurs when a cylinder is not operating at 100-percent capacity because of a weak or failing component or through an error in a sensor reporting to the ECM, which causes the ECM to misinterpret the incoming sensor data. Unfortunately, the causes of random misfires are not always obvious, and the diagnosis can often reach a dead end before discovering the actual cause. Luckily, OBD-II aids in this diagnosis.

OBD-II Misfire Detection

Although the ECM has little to do with correcting misfires, OBD-II gives the ECM the capability to detect misfires, sometimes down to the bank or even the cylinder causing the issue. So how does OBD-II detect a misfiring cylinder? It all comes down to time.

Let's say that I get my exercise by running every day. My route covers eight city blocks for the entire run, and it takes me exactly 8 minutes to make the complete eight-block circuit. Every day I run that same circuit four times. It always takes me 8 minutes to complete the circuit, and because each block is spaced equally, I take 1 minute to run between each block. To keep up on my performance, my lap log looks like this: 8:00 Lap 1, 8:00 Lap 2, 8:05 Lap 3, and 8:00 Lap 4.

Notice that I ran the third lap 5 seconds slower. That tells me that I stumbled or delayed somewhere on that lap momentarily to cause the delay.

I also break down my block-to-block time during an entire route around the eight blocks. That log breaks down like this on my third lap around: 0:60 Block 1, 0:60 Block 2, 0:60 Block 3, 1:05 Block 4, 0:60 Block 5, 0:60 Block 6, 0:60 Block 7, and 0:60 Block 8.

I can see by my logs that I had some sort of delay, somewhere between block number-4 and block number-5. But this was a random occurrence, so I may not be concerned unless it starts to occur every day at the same area.

Now, if I see that I am suddenly running slowly on all four of my laps, then I have a constant delay such as: 8:05 Lap 1, 8:05 Lap 2, 8:05 Lap 3, 8:05 Lap 4.

I know now that I have a problem somewhere. Further study of my block-to-block times shows that all four laps show an added 5 seconds between block number-7 and block number-8. I now can investigate what is constantly occurring between those two blocks.

Just like my workout example, the ECM knows exactly how long a complete crankshaft revolution takes, as well as how long it takes to go between cylinders. It can measure the position of the crankshaft with the crank position sensor and

AUTOMOTIVE DIAGNOSTIC SYSTEMS

CHAPTER 14

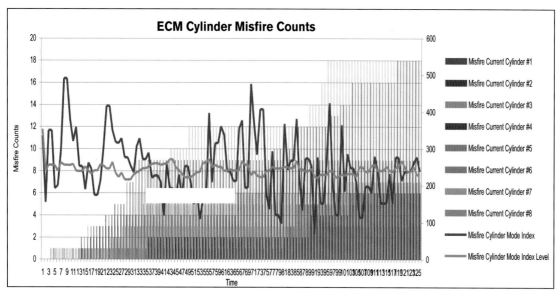

are reset. This amount of time varies by vehicle and manufacturer. If the total amount of misfires exceeds the established limit, a DTC is set and then the MIL is illuminated. This premise works whether the ECM is measuring revolution or cylinder-to-cylinder misfires.

When determining if a revolution misfire is occurring, the ECM determines if the revolution mode index versus the revolution mode index level differ by a set amount. If these two do not agree within the percentage of error, then a misfire count is added to the list. The amount of misfires continues to add to the list until the entire cycle is finished, and then the misfire counts

the reluctor wheel. Because the engine is a dynamic system, the amount of time between events always varies slightly. The ECM references tables to show how much variance it allows before determining there is a misfire occurring.

P030x Misfire DTC

The ECM keeps track of the number of times that it has detected a misfire during a certain amount of time. If the ECM's count of the misfires exceeds a predefined threshold, it sets a DTC and illuminates the MIL on the dashboard. When the ECM can't exactly determine which cylinder is causing the misfire, the ECM sets the P0300 DTC, which indicates that one or more cylinders are randomly misfiring. If the ECM has determined a particular cylinder is misfiring, then it uses the last digit of the P030x code to indicate which cylinder is misfiring. So if cylinder number-3 is misfiring, the ECM sets the P0303 DTC code and illuminates the MIL on the dashboard.

Remember that the misfire checking operation is also one of the system readiness monitors. The ECM must be able to verify that the engine can successfully pass the misfire tests over a defined time period in order to successfully pass the system monitor. Typically, this occurs rather quickly. If the system readiness monitor does not pass, a DTC is often P0300 or a particular cylinder misfire DTC.

Diagnosing a Misfire

Other than a significant mechanical issue with the engine's rotating assembly, the most common causes of an engine misfire are fuel or ignition related: It could be damaged spark plug, damaged spark plug wire, faulty ignition coil, faulty crankshaft position sensor, faulty camshaft position sensor, leaking EGR valve, faulty oxygen sensor, faulty fuel injector, lean or rich fuel condition, or unmetered air (vacuum leak).

Luckily, a DTC is typically set for most of these problems. But, there are times when a bit more diagnosis may be required. These tests typically require a

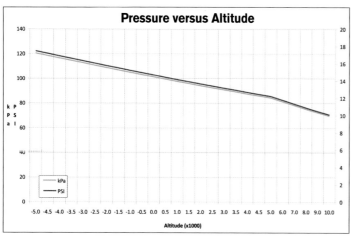

Pressure values are tied directly to the altitude at which they are measured. As the altitude increases, the atmospheric pressure decreases, and vise versa. The ECM needs to determine the altitude based on the static (non-running) engine readings from the MAP sensor, so the timing and fueling maps can be adjusted accordingly.

MISFIRE

few more tools and/or an advanced scan-tool interface.

If a P0300 code is issued, it may mean that the ECM cannot determine which cylinder is misfiring. Some advanced scan tools can test an individual cylinder and have the ECM turn off a single injector to a cylinder. If the idle does not change when that particular cylinder is turned off, that is the suspect cylinder. Another way to test it is to unplug the injector wire for a cylinder and look at the results. I don't recommend pulling a spark plug wire because it could permanently damage the engine.

A cylinder leakdown test determines how well the cylinder is performing. The leakdown test determines if the cylinder can maintain compression within a specified percentage. A weak cylinder could mean anything from a blown head gasket to bad valve to bad rings. A failed leakdown test always means a mechanical rebuild or repair.

A quick test for the health of the fuel system can determine if the fuel injectors are getting adequate fuel flow to feed the engine. When a fuel-related issue causes a misfire, a DTC is usually also set for lean codes. A simple fuel-pressure gauge mounted on the vehicle can determine if the fuel system is delivering fuel pressures within factory specs. It is important to follow all safety instructions when testing any fuel system component.

Two emissions systems can contribute to a misfire situation: the evaporative system and the EGR system. The evaporative system's function is to take vapors coming from the fuel tank and reroute them into the intake to be burned. The valve controls how and when the fumes are introduced. If this valve is not closing, then extra fumes are introduced into the engine. This causes an overly rich condition because the ECM has closed the valve and is not expecting the extra fuel. On the other hand, if the ECM is expecting the extra fuel from the fumes, and the valve is stuck closed, then a lean misfire condition can occur. A DTC is usually established if there is a failure in the evaporative system.

The EGR system also introduces extra combustibles into the cylinder from the exhaust system. The ECM controls the opening and closing of this valve and, if the valve is stuck, then unexpected results and misfires can occur. Again, an accompanying DTC should be stored in the ECM for the EGR system.

Another less common cause for a misfire is a faulty catalytic converter or plugged exhaust system. If the exhaust system is restricted, excessive backpressure can develop. During the exhaust cycle, burned air and fuel must evacuate the cylinder to make room for a fresh air/fuel charge. Failure to evacuate it completely prevents the cylinder from making all the power that it should. This weakens the cylinder and slows the rotational timing down. The ECM recognizes this change as a misfire. Most muffler shops are equipped with equipment to measure exhaust system backpressure and determine if this is causing problems.

Frame-to-Frame Data

The most important part of diagnosing an engine misfire is studying the frame-to-frame data. This data gives a snapshot of the exact conditions of the engine when the misfire occurred. Typically, this shows a lean/rich fueling condition, a high MAP reading indicating a vacuum leak, etc. When a P030x code is encountered, the next action should be pulling the frame-to-frame data for the code.

Blinking MIL

Typically, the ECM sets a P030x DTC when it determines that there is more than a 2-percent variance between cylinder-to-cylinder timing or revolution-to-revolution timing. There are instances where the misfires are much more serious, such as when a cylinder is completely misfiring/dead. If the ECM determines that the variance between cylinders or revolutions exceeds 10 percent of the normal value, the ECM sets a P030x DTC, and begins to rapidly flash the MIL on the dashboard. This typically alerts the driver that a serious misfire is occurring, and the vehicle should be removed from service as quickly as possible for repair.

False Misfire Code

There are times when the ECM sets a P030x DTC and there wasn't a misfire at all. Other powertrain items can create a false misfire code. For instance, if there is an issue with the transmission, such as a clutch that is dragging, or an automatic transmission with a bad torque converter, the component causing the drag or jerking creates a lag in the timing of the cylinders and the ECM measures this. In these cases, the variance can exceed the maximums established by the ECM, and the ECM can set the DTC and illuminate the MIL, even though the engine is operating within the specified parameters. A quick check of the frame-to-frame data shows that everything is operating within specs, so further diagnostics of the drivetrain may be warranted.

Misfires are probably the most frustrating thing to diagnose. Many conditions can cause a misfire, such as a sensor issue or a mechanical issue. When diagnosing a misfire, it is best to map out a consistent plan of attack. Check the more common causes, such as ignition issues and fuel issues, before looking at mechanical issues.

CHAPTER 15

SENSORS

The foundation of the OBD-II system is the sensors and their input into the ECM. Without the sensors, the ECM would be making decisions and trying to diagnose systems with no guidance or feedback. Typically, most issues with modern automobiles deal with faulty sensors delivering improper data to the ECM, or data is not delivered at all. Thankfully, the OBD-II specification has made it possible for the ECM to ensure that the sensors are giving valid data for it to make decisions. The key from a diagnostic standpoint is to at least understand what the key sensors do, what data they deliver to the ECM, and what systems they can effect. This chapter covers the major sensors located on most modern automobiles.

Intake Air Temperature Sensor (IAT)

As discussed in Chapter 11, the air charge density is measured in part with the incoming air temperature in the calculation. The IAT sensor measures this air temperature. The sensor is typically located between the air filter and the throttle body, although some vehicles have it integrated into the MAF sensor. If the MAF sensor has more than three wires, odds are that the IAT is integrated into it.

The IAT sensor is basically an electrical component called a thermistor. The thermistor is a resistor that varies its resistance based on a change in temperature.

$$\Delta R = \Delta T \times k$$

Where: ΔR is the change in resistance, ΔT is the change in temperature and k is the first-order temperature coefficient of resistance.

As the temperature surrounding the IAT changes, the resistance of the sensor changes. This resistance is sent to the ECM where it can equate the resistance from the sensor to a certain temperature.

An IAT can be tested with two methods. The easiest method to verify that the IAT is functioning properly is to observe the readings using the scan tool. By watching the PID associated with the IAT, you can verify that the sensor is

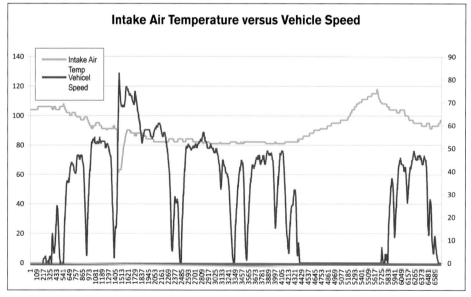

A vehicle at speed has a significant impact on the temperature of the incoming air. The ambient air temperature does not change, per se, but the rushing air around the vehicle's engine bay helps to reduce the heat flowing around the engine, much like blowing air on a hot bowl of soup. As speed decreases, the heated air from around the engine stagnates and is drawn into the intake, thus raising intake air temperatures.

functioning properly. Typically, when an IAT sensor fails, the temperature reports at one of two extremes; very cold (well below 0 degrees F) or extremely hot (well above 200 degrees F). It will be very obvious that the readings are incorrect through the scan tool.

If a scan tool is not available, a VOM can measure the resistance of the sensor by connecting the VOM probes to the two pins on the IAT sensor. Place an IAT sensor in the refrigerator for a few minutes to cool it and take the resistance reading. Then, heat the sensor for a few minutes using a hair dryer or other heat source. Take the resistance reading on the hot sensor and compare it to the cold resistance reading. These two resistance readings should be markedly different. If the two resistance readings are equal or very close to being equal (within a few ohms), then the sensor is most likely bad.

The IAT sensor is a pretty rugged sensor and is located in an area that is not very hostile in terms of heat or chemicals. But there are a few things that can cause an IAT to fail. Typical failures occur in the wiring leading to and from the sensor. Hot underhood temperatures can cause these wires to become brittle. This is such an issue that many dealers sell repair kits, which allow new wires and connectors to be spliced into the harness with minimal effort. Another killer for these sensors is oil contamination. If an aftermarket air filter is over-oiled, there is a chance that the end of the IAT sensor can be contaminated and read incorrectly. When servicing or replacing the IAT sensor, apply dielectric grease to the connector before assembly to prevent water intrusion and corrosion.

When an IAT sensor fails or is reading incorrectly, the vehicle typically displays performance issues and mileage reduction symptoms. Because the IAT is used to establish fueling and timing, an incorrect reading can lead to the ECM applying incorrect ignition timing and incorrect fuel levels. Luckily, a quick test with a scan tool or VOM verifies whether the IAT sensor has failed or not. You don't have to spend a lot of time looking at other sensor inputs or ECM calculations. If the sensor reads incorrectly hot, the vehicle may be hard to start cold, due to incorrect fueling causing a lean condition. If the sensor reads incorrectly cold, then the fueling calculation will be rich, which reduces economy and makes hot starts difficult due to excess fueling on cranking. Since the IAT sensor can also be used by the ECM to command more or less ignition timing, (if spark knock is being experienced) the readings of the IAT sensor should also be checked.

The IAT sensor also controls two other pieces of emissions equipment: the EVAP system and the EGR system. Both of these valves are held closed until a certain, predetermined IAT temperature is reached. Thus, if the IAT is reading incorrectly, both systems may be operating at incorrect times.

The IAT sensor is also paramount to the ECM passing some of its system monitors. A faulty IAT sensor can put the passing of the following monitors in jeopardy:

- Fuel system monitor
- EVAP system monitor
- Misfire monitor

The following DTCs are directly related to a faulty IAT sensor:

Code	Description
P0109	Intake Air Temperature Circuit Malfunction
P0110	Intake Air Temperature Circuit
P0111	Intake Air Temperature Circuit Range/Performance Problem
P0112	Intake Air Temperature Circuit Low Input
P0113	Intake Air Temperature Circuit High Input
P0114	Intake Air Temperature Circuit Intermittent
P0127	Intake Air Temperature Too High
P1111	Intake Air Temperature Sensor Circuit Intermittent High Voltage
P1112	Intake Air Temperature Sensor Circuit Intermittent Low Voltage
P1113	IAT Sensor Open/Short
P1404	IAT - B Circuit Malfunction/ Exhaust Gas Recirculation Closed Position Performance

Engine Coolant Temperature Sensor (ECT)

The ECT has a single job: It measures the temperature of the engine coolant that circulates through the engine's coolant passages. As a result, it effectively measures the temperature of the engine. The sensor is typically located in the engine's cylinder head or in the engine block. Much like the IAT sensor, the ECT sensor is also a thermistor-based sensor, which reports a resistance value to the ECM.

An ECT sensor can be tested with two methods. The easiest method to verify that the ECT sensor is functioning properly is to observe the readings using the scan tool. By watching the PID associated with the ECT, you can verify that the sensor is functioning properly. When an ECT sensor fails, the temperature typically reports at one of the two extremes: very cold (well below 0 degrees F) or extremely hot (well above 200 degrees F). The scan tool makes it very obvious that the readings are incorrect. On a cold engine that has not been run for more than 8 hours, the IAT sensor reading should read very close to the ECT sensor reading. The engine and incoming air should be relatively close to the ambient air temperature surrounding the vehicle.

If a scan tool is not available, the resistance of the ECT sensor can be measured with a VOM. Simply connect the VOM probes to the two pins on the ECT sensor while the sensor is installed on the engine. On a cold engine, remove the connector from the sensor and take a resistance reading. Make sure that the VOM is set on a fairly high resistance value because some sensors can easily show more than 250,000 ohms. Plug the connector back in and allow the engine to come up to temperature (typically around 10 minutes of run time). The connector for the ECT sensor should be carefully disconnected (the engine is hot, be careful not to get burned) and then recheck the resistance. Compare the hot resistance to the cold resistance. If the two resistances are equal, or fairly close, then the ECT sensor should be suspect. If the resistances are far apart, the wiring and connector to the ECT sensor should be checked. When reassembling the connector, use dielectric grease to prevent water intrusion and corrosion.

The ECT sensor is one of the more important sensors because the ECM references its readings when determining fueling and ignition requirements. So when an ECT fails or produces faulty readings, fuel economy and power can be affected. Since the fuel system going into closed loop is determined on engine temperature, a bad ECT prevents the ECM from being able to adjust the fueling, thus preventing the effect on performance, fuel economy, and emissions.

The ECT sensor also determines the idle speed of the engine. On a colder engine, the ECM reads the ECT sensor and uses it to determine the idle speeds required. If the ECT has failed, the engine idles at a much higher engine speed than normal. A more dangerous issue that can occur when an ECT sensor fails is when it fails to the cold side. Based on the readings of the ECT sensor, the ECM controls the engine's cooling fans, and therefore a failure to the cold side never allows the ECM to turn on the fans. This causes the engine to overheat. Because most dashboard engine temperature gauges are directly driven by the value that the ECM reads from the ECT sensor, the driver may never be alerted to the bad ECT sensor until the engine has overheated. As a result, the bad ECT may have stranded the driver and done significant damage to the engine or its components.

The ECT sensor is also paramount to the ECM passing most of its system monitors. A faulty ECT sensor can put the passing of the following monitors in jeopardy:

- Fuel system monitor
- EVAP system monitor
- Misfire monitor
- Heated catalyst monitor
- Catalyst efficiency monitor
- Secondary air system monitor (must get to close-loop operation)
- Oxygen sensor monitor
- Heated oxygen sensor monitor
- EGR system monitor

The following DTCs indirectly relate to a faulty ECT sensor:

Code	Description
P0115	Engine Coolant Temperature Circuit Malfunction
P0116	Engine Coolant Temperature Circuit Range/Performance Problem
P0117	Engine Coolant Temperature Circuit Low Input
P0118	Engine Coolant Temperature Circuit High Input
P0119	Engine Coolant Temperature Circuit Intermittent
P0125	Insufficient Coolant Temperature for Closed-Loop Fuel Control; ECT Excessive Time to Closed-Loop Fuel Control
P0126	Insufficient Coolant Temperature for Stable Operation
P0128	Coolant Thermostat Malfunction
P1114	Engine Coolant Temperature (ECT) Sensor Circuit Intermittent Low Voltage/ IAT - B Circuit Low Input
P1115	Engine Coolant Temperature (ECT) Sensor Circuit Intermittent High Voltage/ IAT - B Circuit High Input
P1116	Engine Coolant sensor out of range/ECT Sensor Out of Self Test Range
P1117	Engine Coolant Sensor Intermittent/ECT Sensor Intermittent
P1798	Coolant Temperature Circuit Malfunction
P1881	Engine Coolant Level Switch Circuit Failure, GEM
P1882	Engine Coolant Level Switch Circuit Short to Ground
P1883	Engine Coolant Level Switch Circuit Failure, GEM
P1884	Engine Coolant Level Lamp Circuit Short to Ground

Throttle Position Sensor (TPS)

The TPS is directly connected to the throttle body and the butterfly inside the throttle body. The TPS measures the angle or percentage that the throttle body's butterfly is opened. When the butterfly is closed, the sensor reads at or near 0 percent and when it is opened, it reads at or near 100 percent. A caveat does exist in some drive-by-wire (electronically controlled) throttle body systems in which the WOT setting is less than 100 percent. Sometimes it's as low as 80 percent, even though the butterfly is still wide open, and ECM's software compensates for this.

The TPS is really a simple potentiometer that varies its resistance internally, depending on the position of the butterfly. Instead of returning a resistance value, the potentiometer acts as a voltage divider and returns a voltage value to the ECM. The lower the voltage value, the more closed the butterfly is. The TPS requires a reference voltage to be sent to the TPS to accurately send a voltage to the ECM. Typically, this reference voltage is +5 volts.

A TPS can be tested using two different methods. Using the scan tool to observe the readings is the easiest method to verify that the TPS sensor is functioning properly. By watching the PID associated with the TPS, you can verify that the sensor is functioning properly. On cable-driven throttle bodies, the position of the TPS can be verified on a non-running vehicle by modulating the accelerator pedal and looking at the readings. On a drive-by-wire throttle body, the engine has to be running to verify the TPS readings. A non-functioning TPS shows no changes while the accelerator pedal is modulated.

If a scan tool is not available, a VOM can measure the TPS's returned voltage value. To take this measurement, you must connect the VOM probes to the ground pin and return pin on the TPS sensor while the accelerator pedal is modulated. It is important to make sure that the transmission is in park or neutral, the brake is set, and the wheels are chocked properly before working on a running engine. The voltage reported on the VOM should vary when the accelerator is modulated, with lesser voltage levels when the throttle body is closed versus higher voltage levels when the throttle body is opened. When reassembling the connector, use dielectric grease to prevent water intrusion and corrosion.

The value of the TPS is important to the ECM because it is used in referencing fueling tables, transmission shift tables, and performance mode calculations. Drivability issues are obvious when a TPS sensor has failed because the ECM tries to understand why extra air is measured in the engine when the TPS is not reporting any activity changes. A TPS that has limited travel prevents the vehicle's transmission from downshifting gears when accelerating, or may prevent the transmission from upshifting during normal driving.

A critical safety factor has been added to vehicles that are equipped with electronic throttle control, or drive-by-wire systems. To prevent runaway engine speeds, if the ECM determines that the TPS has failed, it also determines that it has no control over the throttle body. So it commands the vehicle into a limp mode, which may ignore input from the accelerator pedal. The vehicle still is drivable to the nearest service station, but it only operates at idle speeds. If this occurs, the safest option is to pull the vehicle off the road and have it towed for service.

The TPS is also paramount to the ECM passing some of its system monitors. A faulty TPS can put the passing of the following monitors in jeopardy:

- EVAP system monitor
- Heated catalyst monitor
- Oxygen sensor monitor
- Heated oxygen sensor monitor
- EGR system monitor

The following DTCs indirectly relate to a faulty TPS:

Code	Description
P0120	Throttle Position Sensor/Switch A Circuit Malfunction
P0121	Throttle Position Sensor/Switch A Circuit Range/Performance Problem
P0122	Throttle Position Sensor/Switch A Circuit Low Input
P0123	Throttle Position Sensor/Switch A Circuit High Input
P0124	Throttle Position Sensor/Switch A Circuit Intermittent
P0220	Throttle/Pedal Position Sensor/Switch B Circuit Malfunction
P0221	Throttle/Pedal Position Sensor/Switch B Circuit Range/Performance Problem
P0222	Throttle/Pedal Position Sensor/Switch B Circuit Low Input
P0223	Throttle/Pedal Position Sensor/Switch B Circuit High Input

The throttle position sensor matches the actuation of the accelerator pedal located in the passenger compartment. If engine speed increases as the accelerator is depressed, the TPS sensor should show an increased percentage in proportion to the accelerator pedal.

CHAPTER 15

Code	Description
P0224	Throttle/Pedal Position Sensor/Switch B Circuit Intermittent
P0225	Throttle/Pedal Position Sensor/Switch C Circuit Malfunction
P0226	Throttle/Pedal Position Sensor/Switch C Circuit Range/Performance Problem
P0227	Throttle/Pedal Position Sensor/Switch C Circuit Low Input
P0228	Throttle/Pedal Position Sensor/Switch C Circuit High Input
P0229	Throttle/Pedal Position Sensor/Switch C Circuit Intermittent
P0510	Closed Throttle Position Switch Malfunction
P1120	Throttle Position Sensor Out of Range
P1121	Throttle Position Sensor Circuit Intermittent High Voltage
P1122	Throttle Position Sensor Circuit Intermittent Low Voltage
P1123	Throttle Position Sensor In Range But Higher Than Expected
P1124	Throttle Position Sensor Out of Self Test Range
P1125	Throttle Position Sensor Intermittent
P1126	Throttle Position Sensor Circuit Malfunction
P1167	Invalid Test, Throttle Not Depressed
P1220	Series Throttle Control System Malfunction
P1224	Throttle Position Sensor B Out of Self Test Range
P1511	Idle Switch Circuit Malfunction
P1573	Throttle Position Not Available
P1574	Throttle Position Sensor Disagreement between Sensors
P1575	Pedal Position Out of Self Test Range
P1576	Pedal Position Not Available
P1577	Pedal Position Sensor Disagreement between Sensors

Camshaft Position Sensor (CMP)

The ECM uses the CMP to determine the exact position of the pistons in the engine, as well as the firing position during the Otto Cycle. There are two types of sensors: a Hall Effect sensor and a proximity sensor using a reluctor wheel for positioning. The Hall Effect sensor has two wires coming from the sensor, and the reluctor-style sensor has three wires. Most new vehicles are equipped with the three-wire sensors. (See Chapter 12 for details on how these sensors function.)

If the CMP sensor is defective, the ECM does not know where the engine is in relationship to the Otto Cycle and does not know when to fire the ignition and when to add fuel. As a result, the vehicle is difficult to start (hot or cold), idles rough, and has poor performance, poor fuel economy, and higher emissions output. It may also backfire through the exhaust and intake.

A CMP sensor can be tested by two methods. Using the scan tool to observe the readings is the easiest method to verify that the CMP sensor is functioning properly. By watching the PID associated with the CMP, you can verify that the sensor is functioning properly. The CMP signal shown on the scan tool should be toggling back and forth. Some scan tools show this as a toggling "1" and "0" while others may show it as "open" and "closed." Whatever the method of reporting, the sensor should be toggling between two different states.

If a scan tool is not available, a VOM can still test the CMP but, depending on the style, the test is slightly different. On a two-wire CMP, the VOM should be set to the 5-volt range, and the VOM probes can be back probed (with the connector still connected to the sensor) to the two wires. While the engine is running and the VOM is connected, the voltage should be varying. If the voltage stays constant, the CMP is faulty. On a three-wire system, the first test should be to back probe the reference voltage pin and the ground pin. If this reads close to the battery voltage, the sensor's power section is working properly. The VOM can back probe the ground and signal out pins. While the engine is running and the VOM connected, the voltage should be varying. If the voltage stays constant, the CMP is faulty.

Please note that certain manufacturers require that a dedicated tool be used to locate the position and clocking of the CMP sensor. Failure to properly install the CMP can result in damage to the engine. If in doubt, consult a repair manual or your local auto parts store. Under no circumstances should the CMP sensor be modified.

Furthermore, the ECM uses the CMP sensor and the CKP (crank position sensor) sensor to calculate misfire data. If the ECM sees invalid data coming from the CMP sensor, it may elect to flash the MIL on the dash to indicate a misfire and set misfire DTC codes.

The CMP sensor is also paramount to the ECM passing most of its system monitors. A faulty CMP can threaten the passing of the following sensors:

- Fuel system monitor
- EVAP system monitor
- Misfire monitor
- Heated catalyst monitor
- Catalyst efficiency monitor
- Secondary air system monitor (must get to closed-loop operation)
- Oxygen sensor monitor
- Heated oxygen sensor monitor
- EGR system monitor

The following DTCs are indirectly related to a faulty CMP sensor:

Code	Description
P0340	Camshaft Position Sensor Circuit Malfunction

SENSORS

Code	Description
P0341	Camshaft Position Sensor Circuit Range/Performance
P0342	Camshaft Position Sensor Circuit Low Input
P0343	Camshaft Position Sensor Circuit High Input
P0344	Camshaft Position Sensor Circuit Intermittent
P1174	Cam Sensor Fault
P1175	Cam Control Fault
P1176	Cam Calibration Fault
P1336	Crank/Cam Sensor Range/Performance
P1340	Camshaft Position Sensor B Circuit Malfunction
P1341	Camshaft Position Sensor B Range/Performance
P1345	Cam Position Sensor Circuit Malfunction/Crankshaft Position - Camshaft Position Correlation

Crankshaft Position Sensor (CKP)

The CKP is used by the ECM to determine the exact position of the pistons in the engine, as well as the firing position during the Otto Cycle, by looking at the position of the reluctor wheel attached to the crankshaft. (See Chapter 12 for details on how these sensors function.)

If the CKP sensor is defective, the ECM does not know where the engine is in relationship to the Otto Cycle and does not know when to fire the ignition and when to add fuel. The CKP is the primary sensor to measure engine speed, and most of the ECM's tables reference engine speed. Therefore, with a faulty CKP, the vehicle does not start (hot or cold), or it can intermittently start or not start. An intermittently working CKP also sets misfire codes. Backfires through the exhaust and intake may also be present.

Similar to the CMP sensor, a CKP sensor can be tested by two methods. Using the scan tool to observe the readings is the easiest method to verify that the CKP sensor is functioning properly. By watching the PID associated with the CKP, you can verify that the sensor is functioning properly. The CKP signal shown on the scan tool should be toggling back and forth. Some scan tools show this as a toggling "1" and "0" while others may show it as "open" and "closed." Whatever the method of reporting, the sensor should be toggling between two different states.

If a scan tool is not available, the CKP can still be tested with a VOM, but the VOM cannot interpret all of the information, such as the cylinder position and timing. The first test should be to back probe the reference voltage pin and the ground pin. If this reads close to the battery voltage, the sensor's power section is working properly. The VOM can back probe the ground and signal out pins. While the engine is cranking and the VOM is connected, the voltage should be varying. If the voltage stays constant, the CKP is faulty.

Furthermore, the ECM uses the CKP sensor and the CMP sensor to calculate misfire data. If the ECM sees invalid data coming from the CKP sensor, it may elect to flash the MIL on the dash to indicate a misfire and set misfire DTC codes.

The CKP sensor is also paramount to the ECM passing all of its system monitors because all the monitors reference engine speed. A faulty CKP sensor can put the passing of the following monitors in jeopardy:

- Fuel system monitor
- EVAP system monitor
- Misfire monitor
- Heated catalyst monitor
- Catalyst efficiency monitor
- Secondary air system monitor (must get to closed-loop operation)
- Fuel system monitor
- Oxygen sensor monitor
- Heated oxygen sensor monitor
- EGR system monitor

The following DTCs indirectly relate to a faulty CKP sensor:

Code	Description
P0335	Crankshaft Position Sensor A Circuit Malfunction
P0336	Crankshaft Position Sensor A Circuit Range/Performance
P0337	Crankshaft Position Sensor A Circuit Low Input
P0338	Crankshaft Position Sensor A Circuit High Input
P0339	Crankshaft Position Sensor A Circuit Intermittent
P0385	Crankshaft Position Sensor B Circuit Malfunction
P0386	Crankshaft Position Sensor B Circuit Range/Performance
P0387	Crankshaft Position Sensor B Circuit Low Input
P0388	Crankshaft Position Sensor B Circuit High Input
P0389	Crankshaft Position Sensor B Circuit Intermittent
P1345	Crankshaft Position Sensor Circuit Malfunction- Camshaft Position Correlation

Manifold Absolute Pressure (MAP) Sensor

The process of how vacuum or pressure builds up behind the throttle body during the engine cycle has already been discussed. The MAP sensor is used in the intake manifold to measure the absolute pressure that exists in the manifold. This pressure is then used to determine the density of the incoming airflow, which is the basis for determining the ECM's fueling curves. Some vehicles rely solely on the MAP sensor's input to determine fueling curves (speed-density mode). Other vehicles use the MAF sensor as the primary density measurement sensor with the MAP sensor providing a check on the calculations as well as being used for other load-based ECM calculations. In the world of pressure sensors, there are four main types of sensors: gauge, absolute, vacuum, and differential.

Gauge: A gauge pressure sensor contains a built-in offset measurement. For example, atmospheric pressure is always around 14.7 psi. If you use a tire gauge to measure a flat tire, it reads 0 psi, even though there is always 14.7 psi of air pressure acting on the tire at all times. The tire gauge has been offset by that 14.7 psi to read only pressures that are offset to include atmospheric pressure.

Absolute: An absolute pressure sensor assumes a measurement based on a perfect vacuum pressure of 0 psi. This sensor, if placed perfectly at sea level, measures 14.7 psi. Moving to higher altitudes drops the atmospheric pressure without the sensor compensating for it. Thus, a vehicle at 2,500 feet of altitude has different maximum and minimum MAP sensor values compared to a vehicle at sea level. Some ECMs read the MAP sensor before the engine is running to get a baseline absolute pressure, which it can reference. This may even move the ECM's fueling and timing tables into a high-altitude table to compensate for the difference in operating conditions. The automotive world typically uses the absolute sensor, thus the name manifold *absolute* pressure.

Vacuum: This sensor typically is used to measure pressures that are less than zero atmospheric pressure.

Differential: A differential pressure sensor is actually two pressure sensors built into one component. Typically, there are two pressure hoses coming from the sensor, one that measures pressure in one area, and the other hose measures pressure in another area. The sensor then subtracts one pressure reading from the other and reports the difference between the two pressures. This sensor is very common in industrial applications.

The MAP sensor is used to measure pressure and as a diagnostic aid to the ECM. As stated, the air density can be calculated and compared as a verification test for the MAF sensor. It can also be used to detect a vacuum leak, which shows up as abnormally high pressure readings at given check points. The MAP sensor also checks the EGR system as the

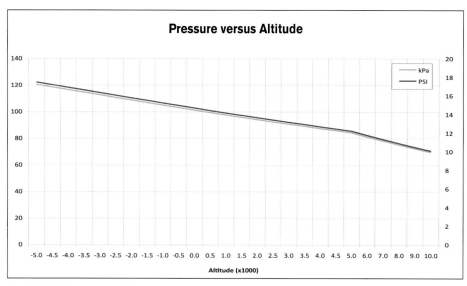

A gauge pressure sensor offsets a base pressure from the measured pressure and returns a reading adjusted to take out the base pressure value from the measurement. In this case, the atmospheric pressure of 101.325 kPa is subtracted from the measured value to give a gauge pressure as shown.

As altitude increases, the atmospheric pressure decreases. Below sea level, atmospheric pressure increases beyond the standard 14.7 psi. Many vehicles are equipped with separate tables in the ECM to allow high-altitude tables to be used. This prevents engine damage and keeps performance levels up while operating at high altitudes.

SENSORS

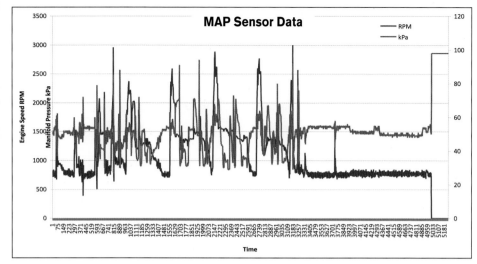

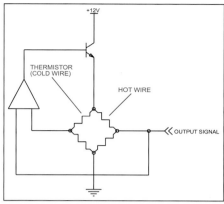

The readings from the MAP sensor often mimic the engine speed as shown. Notice that, at the end of the graph when the engine is turned off, the engine speed goes to zero while the MAP sensor reports the atmospheric pressure (in this case, 98 kPa), which shows that this vehicle was operating slightly above sea level (101.325 kPa).

The basic schematic of a MAF sensor shows the resistor bridge, which consists of a thermistor and a platinum hot wire. The transistor controls the current to the platinum wire so that it is not constantly heated while the vehicle is off.

valve opens and closes. If the MAP sensor does not detect a change when the ECM commands the EGR valve to open, then it sets an EGR DTC and illuminates the MIL.

As with CMP and CKP sensors, a MAP sensor can be tested by two methods. The easiest method to verify that the MAP sensor is functioning properly is to observe the readings using the scan tool. By watching the PID associated with the MAP, you can verify that the sensor is functioning properly. With the engine off, the MAP sensor should be reporting a pressure of around 100 kPa. An idling engine should have pressure readings between 30 and 45 kPa, depending on the vehicle. MAP sensor readings should vary as the vehicle is driven. If the MAP sensor readings do not change, then assume the sensor is faulty.

If a scan tool is not available, a VOM can still test the MAP sensor. The VOM should be set to the 5-volt range, and the VOM probes can be back probed (with the connector still connected to the sensor) to the ground wire and the output signal wire. With the engine off, take a reading on the VOM and write it down as the base atmospheric pressure. While the engine is running and the VOM is connected, take another reading and compare it to the initial reading. The new reading should be less than the first reading by a significant amount. If the voltage stays constant between readings, or does not change much, the MAP sensor is faulty.

The MAP sensor is also paramount to the ECM passing some of its system monitors. A faulty MAP sensor can jeopardize the passing of the following monitors:

- Fuel system monitor
- Misfire monitor
- Heated catalyst monitor
- Catalyst efficiency monitor
- EGR system monitor

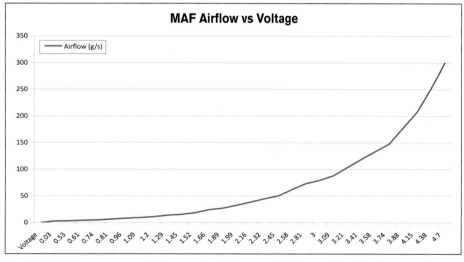

Certain MAF sensors report air mass flow in terms of voltages. These voltages typically range from a low-air-mass flow rate near 0 volts to a maximum-air-mass flow rate just below +5 volts. Typically, the factory chooses an MAF with a scale so that the maximum-air-mass flow rate of the engine does not exceed 75 percent of the scale, which avoids maximizing or exceeding the MAF sensor's capabilities.

AUTOMOTIVE DIAGNOSTIC SYSTEMS

CHAPTER 15

The following DTCs are indirectly related to a faulty MAP sensor:

Code	Description
P0068	MAP/MAF - Throttle Position Correlation
P0069	Manifold Absolute Pressure - Barometric Pressure Correlation
P0105	Manifold Absolute Pressure/Barometric Pressure Circuit Malfunction
P0106	Manifold Absolute Pressure/Barometric Pressure Circuit Range/Performance Problem
P0107	Manifold Absolute Pressure/Barometric Pressure Circuit Low Input
P0108	Manifold Absolute Pressure/Barometric Pressure Circuit High Input
P0109	Manifold Absolute Pressure/Barometric Pressure Circuit Intermittent

Mass Airflow (MAF) Sensor

The MAF sensor is usually the most sensitive and most calibrated sensor in the engine management suite. The MAF sensor is nothing new and has been in use in industrial applications to measure fluid flow for many years. The ECM has to understand the density of the incoming air mass flow, and the MAF sensor is a natural fit for this application. Note that some vehicle manufacturers do not use a MAF sensor on some vehicle models. Instead, those vehicles use a speed/density method of air mass flow calculations.

Several different styles of MAF sensors are available today, but the automotive world typically uses a wire-style sensor, described as a "hot-wire" sensor. Remember, the MAF sensor can respond to changes in air density as well as air volume and send the information to the ECM. The MAF sensor is nothing more than a sensitive piece of test equipment. It suspends in the incoming air stream and simply takes samples thousands of times per second.

The hot-wire MAF sensor is named for the method it uses to measure the incoming airmass flow. Much like the heating elements in a toaster, the hot-wire MAF sensor has one or more wires that are electrically heated to 200 degrees C above the ambient air temperature. Alongside the hot wire is a thermistor, which is known as the cold wire. The cold wire determines the ambient air temperature, which is compared to the temperature of the hot wire. A constant voltage is applied to the hot wire in order

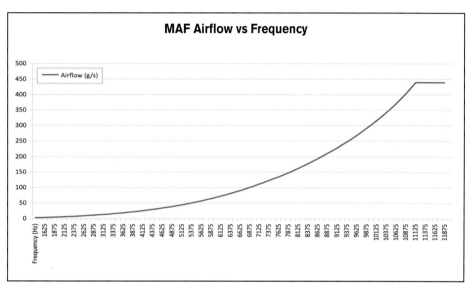

Certain MAF sensors report air mass flow in terms of frequency (timed square wave pulses). These frequencies typically range from a low-air-mass flow rate near 0 Hz to a maximum-air-mass flow rate just below 15,000 Hz. Typically, the factory chooses a MAF scale, so that MAF sensor only measures 75 percent of the maximum-air-mass flow rate of the engine. This prevents maximizing the MAF sensor's capabilities.

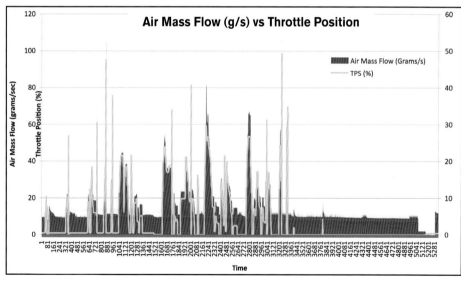

The MAF sensor can be tested by logging its values with a scan tool and modulating the accelerator pedal while the engine is running. The MAF sensor should report higher air mass flow rates when the accelerator is depressed in a proportionate value.

SENSORS

to heat it to the desired temperature. A hot wire has more resistance as temperature increases.

By Ohm's Law, $V = I \times R$ (where V is voltage, I is current, and R is resistance) as resistance increases and voltage remains constant, current drops. As incoming airmass flow cools the hot wire, the resistance decreases and the current increases. The amount of current required to heat the hot wire is proportional to the airmass flow. The MAF sensor has electronics that convert these readings into an output signal for the ECM to interpret. The benefit of this approach is that changes in air temperature and air humidity levels directly affect the amount of current required to heat the wire. Thus, the MAF sensor corrects for a variety of atmospheric conditions.

After the MAF sensor determines the airmass flow, it reports the amount of flow to the ECM as a voltage level or a frequency. If the manufacturer has elected to represent flow in terms of voltage, the ECM has an internal table that correlates that voltage with the airmass flow measurement.

Other manufacturers elect to represent the measurements from the MAF sensor to the ECM as a frequency. The frequency varies up or down depending on the flow measured. The ECM has an internal table that maps frequencies to flow rates.

Because a majority of the fueling calculations rely on the measurement of incoming airmass flow, a faulty MAF sensor throws off the fueling calculations. In the event the MAF sensor is completely non-responsive, the vehicle typically goes into a limp mode so the driver can get the vehicle to the nearest service location. If a MAF sensor is reading incorrectly, fuel mileage decreases, emissions increase, and drivability generally suffers.

Depending on the type of MAF sensor that the vehicle is equipped with, it can be tested in one of two ways. The easiest method to verify that the MAF sensor is functioning properly is to observe the readings using the scan tool. By watching the PID associated with the MAF, you can verify that the sensor is functioning properly. The MAF sensor should report a flow rate that changes as you depress the accelerator pedal and increase RPM. A faulty MAF sensor reports no changes when airflow increases or decreases. A more difficult to detect symptom is when the MAF does not read properly. An experienced technician usually can determine if the flow rates are close to the expected value; you can also consult technical manuals for baseline readings. If neither is practical, you could scan a similarly equipped vehicle, and compare the two scan results for discrepancies.

If the MAF sensor reports the airmass flow rate in terms of voltage, you can use a VOM to measure the voltage from the MAF sensor to the ECM. The VOM should be set to the 5-volt range, and the VOM probes can be back probed (with the connector still connected to the sensor) to the ground wire and the output signal wire. While the engine is running, modulate the throttle and note the voltage reported on the VOM. If the voltage does not change, the MAF sensor may be faulty.

The MAF sensor is a very fragile instrument. Under no circumstances should the wires internal to the sensor be touched. Dirt, oil, and other contaminants can damage the sensor or invalidate its readings. There are several different MAF-sensor-specific cleaners on the market. You simply spray it on the sensor's wires and leave it there to do the cleaning for you. Never spray brake cleaner and other non-MAF cleaners on the MAF sensor; damage may occur. Also, clean the MAF before performing any diagnostics to ensure that contaminants cannot invalidate the data.

The MAF sensor is also paramount to the ECM passing all of its system monitors because each monitor references engine speed. A faulty MAF sensor can threaten the passing of the following monitors:

- Fuel system monitor
- Misfire monitor
- Heated catalyst monitor
- Catalyst efficiency monitor
- Secondary air system monitor (must get to closed-loop operation)
- Oxygen sensor monitor
- Heated oxygen sensor monitor
- EGR system monitor

The following DTCs indirectly relate to a faulty MAP sensor:

Code	Description
P0068	MAP/MAF - Throttle Position Correlation
P0100	Mass or Volume Air Flow Circuit Malfunction
P0101	Mass or Volume Air Flow Circuit Range/Performance Problem
P0102	Mass or Volume Air Flow Circuit Low Input
P0103	Mass or Volume Air Flow Circuit High Input
P0104	Mass or Volume Air Flow Circuit Intermittent

CHAPTER 16

OXYGEN SENSORS

To control the fueling system, the ECM bases its fuel calculations on incoming air mass, engine operating conditions, and altitude. To ensure that the ECM's calculations for fueling are within specification, the entire system runs in closed loop, with the O_2 sensors providing the feedback on the combustion process. The O_2 (or "O2" in OBD codes) sensors are located in the exhaust system at positions that give the sensor the optimal readings of the exhaust stream. Vehicles can have multiple O_2 sensors located in the exhaust system, but there are two main types of O_2 sensors that are based on their location and function.

The first O_2 sensor type is located before the catalytic converters and is known as the up-stream or pre-cat sensor. Its function is to report to the ECM how close the engine is to running at the ideal air/fuel ratio. Typically, these sensors are placed one per engine bank and are located fairly close to where all the primaries of the exhaust manifold come together. This position gives a good sample of the mixture from all the cylinders.

The second O_2 sensor type is located after the catalytic converter (downstream) and is known as the post-cat sensor. Its main function is to evaluate the efficiency of the catalyst and report to the ECM. This sensor is purely designed for emissions quality and has no bearing to the ECM's fuel correction calculations.

Oxygen Sensors Are Consumable

The oxygen sensor is probably *the* most important sensor in the entire closed-loop feedback circuit. So if this sensor fails, or is even degraded, then performance, economy, and emissions can suffer greatly. Few people realize that the oxygen sensor, unlike most other sensors, is a consumable item. This means that the oxygen sensor has a limited lifetime of service and is designed to be replaced when it wears out. This is usually the only sensor on the vehicle that is actually a wear item.

People ask how often the oxygen sensor should be replaced. I've seen oxygen sensors still functioning within specification after 100,000 miles, and I've seen new oxygen sensors fail almost immediately due to other issues with the engine. But, on a normally running, common-service engine, I tell people to replace their pre-cat oxygen sensors when they replace their tires, and this means that oxygen sensors typically should be replaced every 35,000 to 50,000 miles. If the vehicle is not driven on a regular basis or is in a severe-duty application, I recommend replacing sensors every 17,500 to 25,000 miles. The replacement cost of a sensor pales in comparison to the amount of fuel mileage lost due to a degraded or bad sensor. Unfortunately, it may take a while before a bad sensor degrades enough to actually have the ECM set a DTC and illuminate the MIL. That equates to a lot of wasted gasoline.

How an Oxygen Sensor Works

The standard narrow-band oxygen sensor relies on an electrochemical reaction to generate a voltage level, which is reported to the ECM. The electrochemical reaction forms within a solid-state fuel cell known as a Nernst cell, the operation of which is described in the electrochemistry-rooted Nernst Equation. The cell functions at a heat range above 600 degrees F and must maintain that heat electrically or through exhaust temperatures in order to function properly.

The Nernst cell is made up of two plates consisting of zirconium dioxide surrounded by a membrane of platinum. These two plates react to the amount of oxygen in the exhaust stream and generate a voltage using the oxygen ions.

OXYGEN SENSORS

The sensor then uses the outside oxygen content as a baseline to compare it to the exhaust stream. The sampling of the oxygen outside the sensor is actually drawn through the gap between the incoming sensor wires and the insulation on these wires. The gap is very small but sufficient to draw in the sample. This outside sampling method ensures that the air to be sampled isn't fouled or biased in any way. For the sensor to function, it is important to not seal or put any grease on the wires and connector for the oxygen sensor. Those contaminants can block the sampling path and make the sensor's readings invalid.

On a gasoline-powered engine, the stoichiometric air/fuel ratio based on mass is roughly 14.62:1 to 14.7:1. The sensor compares the oxygen levels from the samples outside the sensor to the oxygen levels in the samples in the exhaust stream. When a lean condition occurs in the engine and the air/fuel ratio increases to above 14.7:1, the Nernst cell generates an output voltage of about 200 millivolts (mV) between the two plates. If a rich condition exists (air/fuel ratio is less than stoichiometric), then the Nernst cell generates an output voltage of around 900 mV. When a perfect stoichiometric ratio is reached, the Nernst cell outputs a voltage around 450 mV.

This oxygen sensor is called a "narrow band" sensor because it is only accurate around the stoichiometric air/fuel ratio. As the air/fuel mass ratio moves away from the stoichiometric value, the accuracy of the output voltage versus air/fuel mass ratio decreases. On most production vehicles, this is not an issue. The ECM gives up bandwidth range for greater accuracy around the stoichiometric air/fuel mass ratio because, at this point, the engine operates at its peak efficiency in terms of mileage and emissions. On cold starts, and when at wide-open-throttle events, the air/fuel mass ratio typically is much lower than stoichiometric, but during these times, the ECM typically goes into an open-loop state and does not rely on feedback from the oxygen sensors.

When operating in a typical condition, oxygen sensors generate the voltage

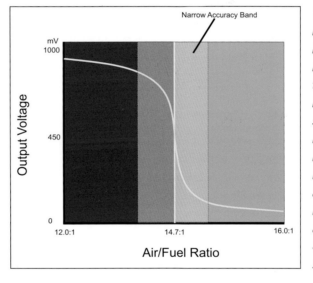

The accuracy of a standard narrow-band oxygen sensor is centered on the stoichiometric air/fuel mass ratio. As the ratio drifts away from the ideal, the accuracy decreases significantly. A higher voltage reported from the sensor indicates a richer air/fuel mass ratio, and a lower voltage indicates a leaner air/fuel mass ratio. Stoichiometric air/fuel mass ratio is around 450 mV on most narrow-band sensors.

There are a few basic parts that make up the narrow-band oxygen sensor. Exhaust gases pass through the tip of the sensor where they interact with the zirconium dioxide element. This interaction passes a voltage to the surrounding platinum electrodes, which connect to the voltage-return side of the sensor. The ground for the sensor is picked up through the connection of the sensor to the exhaust pipe.

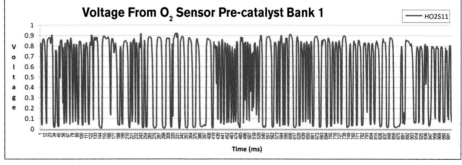

An oxygen sensor, located pre-catalyst in a heated running condition, exhibits an oscillating voltage graph. Most scanners cannot read the data because the switching frequency of these numbers is too fast for the scanner. The best scanners for diagnosing oxygen sensor issues can output the data to be post-processed or provide a graphical view of the data.

AUTOMOTIVE DIAGNOSTIC SYSTEMS

CHAPTER 16

that swings back and forth from rich to lean in a sinusoidal pattern. The voltage should vary from a low of around 200 mV to a high of around 900 mV. The pattern varies slightly due to the dynamics of the exhaust system and sample rates, but after a healthy sensor has warmed up, it should always have a form of oscillating pattern.

Adequate temperature is very important for proper oxygen sensor operation. Almost all modern-day oxygen sensors contain a heating element to ensure that the sensor reaches its peak operating temperature quickly and maintains it properly. Most of the sensors work best at around 600 degrees F, but manufacturers can vary that depending on oxygen sensor position and manufacturing techniques of the sensor itself. If a sensor cannot maintain the proper temperatures, it most likely won't produce correct voltage values to inform the ECM of the air/fuel mass ratio.

In a startup scenario, the oxygen sensor flatlines its voltage right around 450 mV. As the sensor comes up to operating temperature, it slowly starts to oscillate. Until it comes up to temperature, the voltage values that the sensor is reporting cannot be considered legitimate. Thus, most ECMs wait for a certain amount of time to pass in order to verify that the oxygen sensor's voltages are switching.

Diagnosing an Oxygen Sensor

A properly functioning narrow-band oxygen sensor oscillates as previously discussed. To test it with a graphing scanner, two points of tests must be performed to ensure that the data that the sensor is reporting is within norm. You also need to ensure that another engine issue does not bias the tests.

While graphing the pre-catalyst oxygen sensor, the engine should be brought to operating temperature and idled for approximately 3 to 4 minutes. The throttle should be momentarily depressed to a WOT position. The oxygen sensor should change readings from idle to values between 100 mV and 900 mV. If the sensor readings don't change, then it could be an air/fuel delivery issue. Next, bring up the engine speed and hold it between 2,300 and 2,600 rpm. The voltage levels from the oxygen sensor should change and start to oscillate approximately every 1 second. As long as the voltage levels are changing during these tests, the oxygen sensor is most likely working properly.

If the signal changes, but remains low, getting only as high as 600 to 700

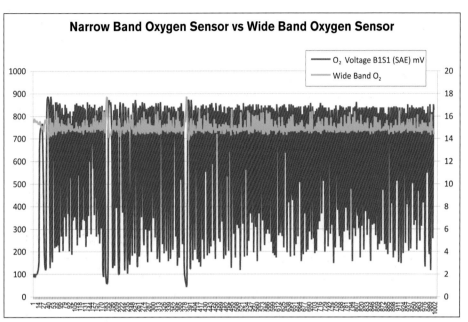

The voltage values returned from a narrow-band oxygen sensor are plotted against a wide-band oxygen sensor, showing the actual air/fuel mass ratio. The actual air/fuel mass ratio is very close to stoichiometric on the wide-band sensor, but the accuracy of the narrow-band sensor around that value shows significant voltage swings. This is the benefit of having an accurate (albeit narrow range) sensor.

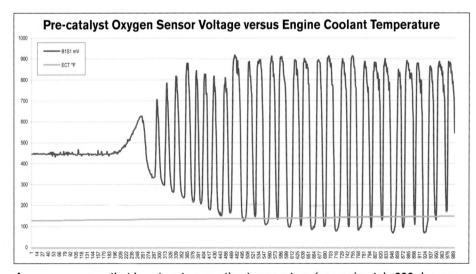

An oxygen sensor that is not up to operating temperature (approximately 600 degrees F) does not report correct air/fuel mass ratios. As the sensor's internal temperature rises, the voltage output values start to oscillate in the proper manner.

OXYGEN SENSORS

mV or less, the air/fuel mass ratio is lean. Poor fuel delivery, a vacuum leak, or other factors could cause this condition. If the sensor values oscillate, but remain high and never go below 450 mV, then the engine is running in a rich condition. Both the lean condition and rich condition are typically the fault of a mechanical issue or a bad sensor. (You should study further scan tool information on the other fuel-dependant sensors.)

A quick method to determine if a sensor is reading accurately is to display the graphical data of the oxygen sensor and then force a lean, followed by a rich, condition on the engine. Unplugging a vacuum line after the MAF sensor (downstream) can introduce a lean condition, which allows unmetered air to be drawn in without the ECM adding fuel for it. The oxygen sensor should immediately report a leaner reading. To create a rich condition, you can introduce propane to the engine through a port or vacuum line. When added, the propane registers as excess fuel at the oxygen sensor, and the voltage increases proportionately.

Oxygen Sensor Codes

Luckily, the ECM keeps a pretty tight watch over the performance of the oxygen sensors. Through OBD-II specs, these provide the main feedback for the emissions and mileage calculations. And when an oxygen sensor goes bad, the ECM typically establishes the corresponding DTC and illuminates the MIL.

The oxygen sensor is also paramount to the ECM passing most of its system monitors. A faulty pre-catalytic oxygen sensor can threaten the passing of the following monitors:

- Fuel system monitor
- EVAP system monitor
- Misfire monitor
- Heated catalyst monitor
- Catalyst efficiency monitor
- Secondary air system monitor (must get to closed-loop operation)
- Oxygen sensor monitor
- Heated oxygen sensor monitor
- EGR system monitor

The following DTCs are indirectly related to a faulty ECT sensor:

Code	Problem
P0030	HO2S Heater Control Circuit (Bank 1 Sensor 1)
P0031	HO2S Heater Control Circuit Low (Bank 1 Sensor 1)
P0032	HO2S Heater Control Circuit High (Bank 1 Sensor 1)
P0036	HO2S Heater Control Circuit (Bank 1 Sensor 2)
P0037	HO2S Heater Control Circuit Low (Bank 1 Sensor 2)
P0038	HO2S Heater Control Circuit High (Bank 1 Sensor 2)
P0040	Upstream Oxygen Sensors Swapped From Bank to Bank
P0041	Downstream Oxygen Sensors Swapped From Bank to Bank
P0042	HO2S Heater Control Circuit (Bank 1 Sensor 3)
P0043	HO2S Heater Control Circuit Low (Bank 1 Sensor 3)
P0044	HO2S Heater Control Circuit High (Bank 1 Sensor 3)
P0050	HO2S Heater Control Circuit (Bank 2 Sensor 1)
P0051	HO2S Heater Control Circuit Low (Bank 2 Sensor 1)
P0052	HO2S Heater Control Circuit High (Bank 2 Sensor 1)
P0053	HO2S Heater Resistance (Bank 1 Sensor 1)
P0054	HO2S Heater Resistance (Bank 1 Sensor 2)
P0055	HO2S Heater Resistance (Bank 1 Sensor 3)
P0056	HO2S Heater Control Circuit (Bank 2 Sensor 2)
P0057	HO2S Heater Control Circuit Low (Bank 2 Sensor 2)
P0058	HO2S Heater Control Circuit High (Bank 2 Sensor 2)
P0059	HO2S Heater Resistance (Bank 2 Sensor 1)
P0060	HO2S Heater Resistance (Bank 2 Sensor 2)
P0061	HO2S Heater Resistance (Bank 2 Sensor 3)
P0062	HO2S Heater Control Circuit (Bank 2 Sensor 3)
P0063	HO2S Heater Control Circuit Low (Bank 2 Sensor 3)
P0064	HO2S Heater Control Circuit High (Bank 2 Sensor 3)
P0130	O2 Sensor Circuit Malfunction (Bank 1 Sensor 1)
P0131	O2 Sensor Circuit Low Voltage (Bank 1 Sensor I)
P0132	O2 Sensor Circuit High Voltage (Bank 1 Sensor 1)
P0133	O2 Sensor Circuit Slow Response (Bank 1 Sensor 1)
P0134	O2 Sensor Circuit No Activity Detected (Bank 1 Sensor 1)
P0135	O2 Sensor Heater Circuit Malfunction (Bank 1 Sensor 1)
P0136	O2 Sensor Circuit Malfunction (Bank 1 Sensor 2)
P0137	O2 Sensor Circuit Low Voltage (Bank 1 Sensor 2)
P0138	O2 Sensor Circuit High Voltage (Bank 1 Sensor 2)
P0139	O2 Sensor Circuit Slow Response (Bank 1 Sensor 2)
P0140	O2 Sensor Circuit No Activity Detected (Bank 1 Sensor 2)
P0141	O2 Sensor Heater Circuit Malfunction (Bank 1 Sensor 2)
P0142	O2 Sensor Circuit Malfunction (Bank 1 Sensor 3)
P0143	O2 Sensor Circuit Low Voltage (Bank 1 Sensor 3)
P0144	O2 Sensor Circuit High Voltage (Bank 1 Sensor 3)
P0145	O2 Sensor Circuit Slow Response (Bank 1 Sensor 3)

Code	Description
P0146	O2 Sensor Circuit No Activity Detected (Bank 1 Sensor 3)
P0147	O2 Sensor Heater Circuit Malfunction (Bank 1 Sensor 3)
P0150	O2 Sensor Circuit Malfunction (Bank 2 Sensor 1)
P0151	O2 Sensor Circuit Low Voltage (Bank 2 Sensor 1)
P0152	O2 Sensor Circuit High Voltage (Bank 2 Sensor 1)
P0153	O2 Sensor Circuit Slow Response (Bank 2 Sensor 1)
P0154	O2 Sensor Circuit No Activity Detected (Bank 2 Sensor 1)
P0155	O2 Sensor Heater Circuit Malfunction (Bank 2 Sensor 1)
P0156	O2 Sensor Circuit Malfunction (Bank 2 Sensor 2)
P0157	O2 Sensor Circuit Low Voltage (Bank 2 Sensor 2)
P0158	O2 Sensor Circuit High Voltage (Bank 2 Sensor 2)
P0159	O2 Sensor Circuit Slow Response (Bank 2 Sensor 2)
P0160	O2 Sensor Circuit No Activity Detected (Bank 2 Sensor 2)
P0161	O2 Sensor Heater Circuit Malfunction (Bank 2 Sensor 2)
P0162	O2 Sensor Circuit Malfunction (Bank 2 Sensor 3)
P0163	O2 Sensor Circuit Low Voltage (Bank 2 Sensor 3)
P0164	O2 Sensor Circuit High Voltage (Bank 2 Sensor 3)
P0165	O2 Sensor Circuit Slow Response (Bank 2 Sensor 3)
P0166	O2 Sensor Circuit No Activity Detected (Bank 2 Sensor 3)
P0167	O2 Sensor Heater Circuit Malfunction (Bank 2 Sensor 3)
P0170	Fuel Trim Malfunction (Bank 1)
P0171	System Too Lean (Bank 1)
P0172	System Too Rich (Bank 1)
P0173	Fuel Trim Malfunction (Bank 2)
P0174	System Too Lean (Bank 2)
P0175	System Too Rich (Bank 2)
P0420	Catalyst System Efficiency Below Threshold (Bank 1)
P0421	Warm Up Catalyst Efficiency Below Threshold (Bank 1)
P0422	Main Catalyst Efficiency Below Threshold (Bank 1)
P0423	Heated Catalyst Efficiency Below Threshold (Bank 1)
P0424	Heated Catalyst Temperature Below Threshold (Bank 1)
P0430	Catalyst System Efficiency Below Threshold (Bank 2)
P0431	Warm Up Catalyst Efficiency Below Threshold (Bank 2)
P0432	Main Catalyst Efficiency Below Threshold (Bank 2)
P0433	Heated Catalyst Efficiency Below Threshold (Bank 2)
P0434	Heated Catalyst Temperature Below Threshold (Bank 2)
P1127	Exhaust Not Warm, Downstream O2 Sensor
P1128	Upstream Heated O2 Sensors Swapped
P1129	Downstream Heated O2 Sensors Swapped
P1130	Lack of HO2S Switch - Adaptive Fuel At Limit
P1131	Lack of HO2S Switch - Sensor Indicates Lean
P1132	Lack of HO2S Switch - Sensor Indicates Rich
P1133	HO2S Insufficient Switching Sensor 1
P1134	HO2S Transition Time Ratio Sensor 1
P1137	Lack of HO2S Switch - Sensor Indicates Lean
P1138	Lack of HO2S21 Switch - Sensor Indicates Rich
P1150	Lack of HO2S21 Switch - Adaptive Fuel At Limit
P1151	Lack of HO2S21 Switch - Sensor Indicates Lean
P1152	Lack of HO2S21 Switch - Sensor Indicates Rich
P1153	Bank 2 Fuel Control Shifted Lean
P1154	Bank 2 Fuel Control Shifted Rich
P1157	Lack Of HO2S22 Switch - Sensor Indicates Lean
P1158	Lack Of HO2S22 Switch - Sensor Indicates Rich

What Causes Oxygen Sensor Failure?

Oxygen sensors operate in a demanding environment and therefore are pretty tough sensors. But, as previously stated, they are considered a consumable sensor that eventually wears out and requires replacement. There are several factors that can accelerate their replacement schedule. These factors include:

- Running the incorrect fuel type in the vehicle. Leaded gasoline, such as leaded race gas, shortens the life of an oxygen sensor to as short as a matter of hours. A sensor that has been running in leaded gasoline typically has a rusty color on the tip of the sensor.
- Engine oil ruins a sensor quickly by fouling the tip of the sensor, so that it can no longer sample the incoming exhaust stream. If an engine is burning oil through a mechanical issue or a PCV problem, the oxygen sensor's performance degrades rapidly.
- Excessive grease, dirt, or other contaminants built up on the outside of the sensor wires reduces or eliminates the capability for the sensor to draw in outside oxygen for sampling. Sometimes the wires can be cleaned but, due to the miniscule size of the passages, they are difficult to get completely free of contaminants.
- If engine repairs are made and sealants are used, the sealants must be labeled as "sensor friendly." Many sealants contain chemicals, which will foul or damage the sensor if the chemicals can reach the exhaust stream.

OXYGEN SENSORS

- If a head gasket or intake gasket is leaking and antifreeze is being burned in the combustion, the byproduct of the burned antifreeze is poisonous to the oxygen sensor and ruins it quickly.
- Overly rich conditions can foul the tip of the oxygen sensor with black soot, rendering the heater circuit or the sensor itself useless. Occasionally, the tip can be burned clean with a propane torch, but usually, the sensor has to be replaced.

Replacing an Oxygen Sensor

An at-home mechanic often removes a brand-new oxygen sensor from the box and improperly installs it. This mistake ruins it, or at least shortens its life. When replacing the oxygen sensor, here are some tips to gain the most life possible from it:

- Always use an oxygen sensor wrench or socket when removing and installing the sensor. These tools are designed to apply even torque to the sensor. Under no circumstances should an adjustable wrench be used on an oxygen sensor.
- In order to facilitate easy removal of a sensor at a later time, apply anti-seize compound to the threads prior to installation. Take care to only apply a small dab to the threads, and don't get any compound on the tip of the sensor.
- The wires in the sensor are delicate and can be broken or pinched. Install the sensor prior to connecting the wires to the harness. Also take care when routing the wires. Exhaust systems are extremely hot and can quickly melt oxygen sensor wires.
- Oxygen sensors typically come with a plastic cover over the tip. Keep this cover on the sensor until you are ready to thread it into the exhaust system. This helps prevent contaminants from entering the sensor.
- Remember that oxygen sensors operate at 600 degrees F or more. Even if the vehicle is not running, if the ignition is on, the sensor's heaters will most likely be engaged and heating the sensor. Prior to removing the sensor, turn off the engine and let it cool for an adequate amount of time.
- When changing exhaust systems or exhaust pipes, keep in mind that the pipes are often coated with light oil when being bent. This oil is usually not cleaned off prior to installation, burns off and can harm an oxygen sensor. If the exhaust components prior to the oxygen sensor are replaced, run the engine with the exhaust sensor removed for 10 minutes to burn off the excess manufacturing oil.

GLOSSARY

ABDC: After Bottom Dead Center
ALCL: Assembly Line Communications Link
ALDL: Assembly Line Diagnostics Link
AIR: Secondary Air Injection
ARB: Air Resources Board
ATDC: After Top Dead Center
BBDC: Before Bottom Dead Center
BDC: Bottom Dead Center
BTDC: Before Top Dead Center
CAFE: Corporate Average Fuel Economy
CAN: Controller Area Network
CARB: California Air Resources Board
CCM: Comprehensive Component Monitoring
CEL: Check Engine Light
CKP: Crankshaft Position Sensor
CMP: Camshaft Position Sensor
CNG: Compressed Natural Gas
CRC: Cyclic Redundancy Check
DLC: Data Link Connector
DTC: Diagnostic Trouble Code
ECM: Engine Control Module
ECT: Engine Coolant Temperature
EEPROM: Electrically Erasable Programmable
    Read-Only Memory
EGR: Exhaust Gas Recirculation
EO: Executive Order
EOD: End Of Data
EOF: End Of File
EPA: Environmental Protection Agency
HVAC: Heating, Venting, and Air Conditioning

IAT: Intake Air Temperature
IFR: In-Frame Response
IM: Inspection/Maintenance
ISO: International Organization for Standards
LPG: Liquefied Petroleum Gas
LTFT: Long-Term Fuel Trim
MAF: Mass Airflow Sensor
MAP: Manifold Air Pressure
MIL: Malfunction Indicator Light
NASTF: National Automotive Service Task Force
NHTSA: National Highway Traffic Safety Administration
OBD: On-Board Diagnostics
OEM: Original Equipment Manufacturer
OSI: Open Systems Interconnection
PID: Parameter Identification
PWM: Pulse Width Modulated
RPM: Revolutions Per Minute
SAE: Society of Automobile Engineers
SOF: Start Of Frame
SRM: System Readiness Monitors
SRT: System Readiness Tests
STFT: Short-Term Fuel Trim
TDC: Top Dead Center
TPS: Throttle Position Sensor
TSB: Technician Service Bulletin
VOC: Volatile Organic Compounds
VOM: Volt Ohm Meter
VPW: Variable Pulse Width
VSS: Vehicle Speed Sensor
WOT: Wide-Open Throttle

APPENDIX A

USING A VOLT OHM METER

A VOM is one of the most useful tools when diagnosing any electrical issue, whether it is on a vehicle or in the home. The VOM is also known as a multimeter because of its multifunctional capability to measure voltage, resistance, and amperage. The VOM should be one of the first testing devices that a mechanic should invest in—maybe even before purchasing a scan tool, because many automotive shops read trouble codes for free. But it is also very important to completely understand how to operate a VOM; there is an element of danger anytime you test and diagnose electrically powered systems. Failure to operate the VOM properly can result in personal injury and/or damage to the vehicle. If you are uncomfortable working with electrical systems and a VOM, ask for help from someone who is comfortable using a VOM.

A VOM can cost from $25 to more than $1,000. But, the price of the meter doesn't always reflect its usefulness. Many times, the simple $25 VOM performs satisfactorily for most home enthusiasts. The upper-end VOMs usually have extra functionality for lab use and are much more accurate. For most automotive applications, the accuracy of an inexpensive VOM is more than enough to help diagnose a system.

Features

Here are a few key features to look for when purchasing a VOM:

Auto Ranging

Newer VOMs automatically change the scale of the readings based on the readings that they are taking. So the same voltage settings read 5 volts DC and without changing the settings read 120 volts AC. On an older VOM without auto ranging, if the meter was set at 10-to-100-volt readings, and the meter was plugged into a standard house outlet at 120 volts, the meter would stop reading at 100 volts or show that it is out of range. A more important auto ranging feature comes when resistance is being measured. Without having any knowledge of the resistance of a circuit, without auto ranging, you would have to cycle through the

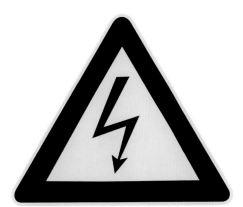

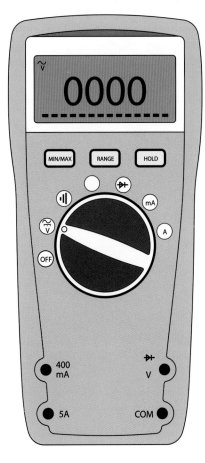

AUTOMOTIVE DIAGNOSTIC SYSTEMS 117

ranges until it did not peg the meter. A meter that auto ranges saves a lot of time and frustration while diagnosing.

Continuity Check

Along with being able to read resistance, some VOMs are capable of checking for continuity and giving an audible beep when continuity is detected. When searching for broken wires or shorts, having the meter beep without having to look at it makes diagnosing these issues much easier.

Diode Check

A VOM that can check for a blown diode is extremely handy. There are diodes throughout an automobile's electrical system. Knowledge of a blown diode can save a lot of time and trouble.

Probes

The VOM should include a good, solid set of probes with banana clips on the ends. Make sure that the probe connections are standard so you can easily replace them. Make sure that the gauge (thickness) of the probe wires is also adequate. As a general rule, the probe wires should be at least 18 gauge in automotive applications.

Replacement Fuses

Everyone makes mistakes when diagnosing electrical systems. Usually, these mistakes are shorting power to ground, which can damage the meter. Good meters have an easily replaceable fuse, which protects the user and meter. Keep a few extra fuses on hand.

VOM Connections

Most VOMs have at least two connections, but some may have three or more. All VOMs have a common plug connection for the black probe. The other connections depend on the capabilities of the VOM. At a minimum, a connection is provided for measuring voltage and resistance. On modern-day VOMs, this connection can measure alternating current (AC) and direct current (DC) and voltage and resistance. The connections are labeled, and you should confirm this before plugging the probes in and testing the circuit. You can also use this connection to measure resistance and conductivity, and to perform diode checks on some VOMs. If the VOM has the capability to measure current, a separate connection is usually supplied. The current capabilities (maximum current in milliamps or amps) are labeled next to the connection point. It is critical to not exceed the meter's rated capabilities.

In the example below, the meter has two connections for measuring amperage. The first connection is capable of measuring up to and including 400 milliamps. Using this connection, the VOM mimics a milliamp meter. The second connection shows the maximum meter capability of 5 amps. In this mode, the VOM mimics an amp meter. Although the 5-amp connection can be used to measure from 0 to 5 amps, if the current is below 400 milliamps, it is advisable to use the 400-milliamp connection due to its increased resolution and accuracy.

Measuring Voltage with a VOM

The VOM is capable of measuring AC and DC voltage. If the VOM is auto detecting and auto ranging, simply set the dial to "VOLTAGE" to measure voltage. If the meter requires the type of voltage to be set, then rotate the dial to the proper voltage type, AC or DC. Automotive applications always use DC voltage. If the VOM is not auto ranging, then the range of the measured voltage must also be set.

Connect the black probe to the "COM," or common connection. Think of this as ground in electrical circuits. Connect the red probe to the "VOLTAGE" connection. In the example, the voltage of a battery is being measured. The red probe is connected to the positive terminal of the battery, and the black probe is connected to the negative terminal of the battery. If these connections are reversed, the voltage on the VOM reads negative, or, in this example, it would read –12.10 volts. The meter will not be damaged if the polarity flipped in this instance. Any voltage can be measured throughout the vehicle as long as a good ground is connected to the black probe. When measuring voltage, make sure that the probes do not touch anything other than the

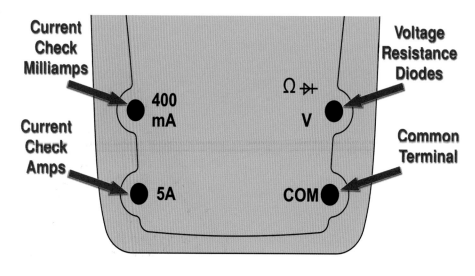

USING A VOLT OHM METER

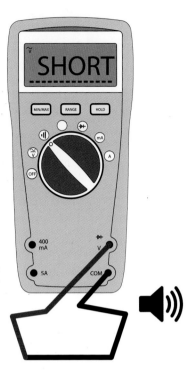

area to be tested. If one touches ground, a connection is made, possibly causing harm to the technician, the meter, or the automobile.

Continuity Tests

The continuity test on the VOM (if so equipped) is an extremely handy test when looking for shorts or opens in wiring. A "short" means that the circuit is completed so that voltage and current flows continuously through it. Sometimes this is desirable, and other times not. If the battery is shorted to the chassis, you can receive a shock each time you touch the chassis and complete the circuit. An "open" is just the opposite; there is no connection between the points. If you test a spark plug wire and no continuity is shown, then it is open and likely a bad spark plug wire. A VOM with continuity-test capability usually alerts with an audio indication (a beep, a chime, a tone, etc.) when a short has been probed. This audio acknowledgement allows you to continually probe without having to look at the meter.

To perform a continuity test using a VOM, set the dial to the continuity test. Connect the black probe to the COM or common connection, and connect the red probe to the "Ω" or "CONT" connection. Then, touch the probes to the two ends of where you are checking for a short or open in the circuit. The audio feedback, as well as a visual report on the meter, alerts the status. When probing a line that has voltage on it, make sure that neither probe goes to ground. Shorts can cause physical damage to the user, the system, and the meter.

Measuring Current with a VOM

The VOM is capable of becoming an amp meter or milliamp meter to measure the current in a circuit. The difference in measuring current, versus any of the other measurements with the VOM, is that you place the VOM in series in the circuit. Thus, you must break the circuit and insert the VOM into the circuit so that the current goes continuously through the VOM. If you place the VOM in parallel instead of in series, then the readings will be inaccurate.

Before connecting the VOM, turn the circuit off. Next, connect the black probe to the COM or common connection, and connect the red probe to the desired amperage range connection. If you are not certain of the amount of current in the circuit, then it's best to hook the red probe to the meter's highest current range to avoid damaging the meter. If you determine that the current is lower than the meter can detect, you can move the probe to obtain more resolution and accuracy. When the probes are in place, you can power-on the circuit and measure the current.

APPENDIX A

Measuring Toggling Signals

On most digital computer-controlled systems, signal lines are changing from high to low voltage in order to pass data from the sender to the receiver. These oscillating diagnostic signals occur in many places on the automobile. Examples include fuel-injector control lines, speed sensors, mass airflow signals, and many more. The issue is that these voltages change so fast that a normal VOM cannot show these changes quickly enough or give information that you can use. At this point, there are really two choices to diagnose these types of issues: a graphing VOM or an oscilloscope.

I use a multi-channel oscilloscope because I need it for my engineering design work, but the price of these scopes starts at $2,500 and goes up quickly. But you can find older oscilloscopes on online auctions and at schools for a good price. Graphing VOMs also perform the basic operations of an oscilloscope at a significantly reduced cost. You can find these priced around $600 and up. A fairly new product that allows a laptop or personal computer to be used as a digital oscilloscope is available. Prices on these are as low as $250, and they are suitable and heavy-duty enough for automotive diagnostic.

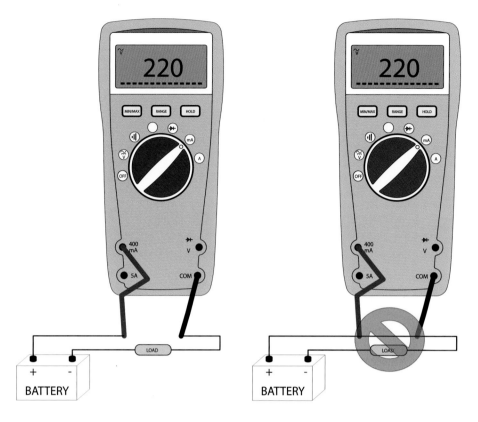

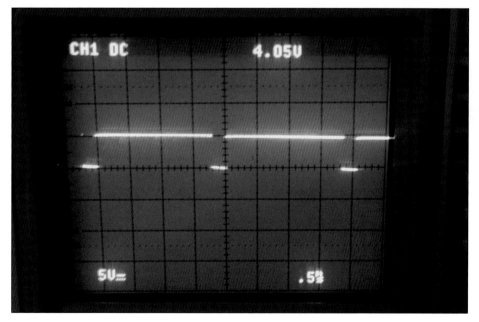

APPENDIX B

Generic OBD-II DTC Codes

P0001 to P0099: Fuel Metering, Air Metering and Auxiliary Emission Controls

P0001	Fuel Volume Regulator Control Circuit/Open
P0002	Fuel Volume Regulator Control Circuit Range/Performance
P0003	Fuel Volume Regulator Control Circuit Low
P0004	Fuel Volume Regulator Control Circuit High
P0005	Fuel Shutoff Valve "A" Control Circuit/Open
P0006	Fuel Shutoff Valve "A" Control Circuit Low
P0007	Fuel Shutoff Valve "A" Control Circuit High
P0008	Engine Position System Performance
P0009	Engine Position System Performance
P0010	"A" Camshaft Position Actuator Circuit
P0011	"A" Camshaft Position - Timing Over-Advanced or System Performance
P0012	"A" Camshaft Position - Timing Over-Retarded
P0013	"B" Camshaft Position - Actuator Circuit
P0014	"B" Camshaft Position - Timing Over-Advanced or System Performance
P0015	"B" Camshaft Position - Timing Over-Retarded
P0016	Crankshaft Position - Camshaft Position Correlation
P0017	Crankshaft Position - Camshaft Position Correlation
P0018	Crankshaft Position - Camshaft Position Correlation
P0019	Crankshaft Position - Camshaft Position Correlation
P0020	"A" Camshaft Position Actuator Circuit
P0021	"A" Camshaft Position - Timing Over-Advanced or System Performance
P0022	"A" Camshaft Position - Timing Over-Retarded
P0023	"B" Camshaft Position - Actuator Circuit
P0024	"B" Camshaft Position - Timing Over-Advanced or System Performance
P0025	"B" Camshaft Position - Timing Over-Retarded
P0026	Intake Valve Control Solenoid Circuit Range/Performance
P0027	Exhaust Valve Control Solenoid Circuit Range/Performance
P0028	Intake Valve Control Solenoid Circuit Range/Performance
P0029	Exhaust Valve Control Solenoid Circuit Range/Performance
P0030	HO2S Heater Control Circuit
P0031	HO2S Heater Control Circuit Low
P0032	HO2S Heater Control Circuit High
P0033	Turbo Charger Bypass Valve Control Circuit
P0034	Turbo Charger Bypass Valve Control Circuit Low
P0035	Turbo Charger Bypass Valve Control Circuit High
P0036	HO2S Heater Control Circuit
P0037	HO2S Heater Control Circuit Low
P0038	HO2S Heater Control Circuit High
P0039	Turbo/Super Charger Bypass Valve Control Circuit Range/Performance
P0040	O2 Sensor Signals Swapped Bank 1 Sensor 1/ Bank 2 Sensor 1
P0041	O2 Sensor Signals Swapped Bank 1 Sensor 2/ Bank 2 Sensor 2
P0042	HO2S Heater Control Circuit
P0043	HO2S Heater Control Circuit Low
P0044	HO2S Heater Control Circuit High
P0045	Turbo/Super Charger Boost Control Solenoid Circuit/Open
P0046	Turbo/Super Charger Boost Control Solenoid Circuit Range/Performance
P0047	Turbo/Super Charger Boost Control Solenoid Circuit Low
P0048	Turbo/Super Charger Boost Control Solenoid Circuit High
P0049	Turbo/Super Charger Turbine Overspeed
P0050	HO2S Heater Control Circuit
P0051	HO2S Heater Control Circuit Low
P0052	HO2S Heater Control Circuit High
P0053	HO2S Heater Resistance
P0054	HO2S Heater Resistance
P0055	HO2S Heater Resistance
P0056	HO2S Heater Control Circuit
P0057	HO2S Heater Control Circuit Low
P0058	HO2S Heater Control Circuit High
P0059	HO2S Heater Resistance
P0060	HO2S Heater Resistance
P0061	HO2S Heater Resistance
P0062	HO2S Heater Control Circuit

Code	Description
P0063	HO2S Heater Control Circuit Low
P0064	HO2S Heater Control Circuit High
P0065	Air Assisted Injector Control Range/Performance
P0066	Air Assisted Injector Control Circuit or Circuit Low
P0067	Air Assisted Injector Control Circuit High
P0068	MAP/MAF - Throttle Position Correlation
P0069	Manifold Absolute Pressure - Barometric Pressure Correlation
P0070	Ambient Air Temperature Sensor Circuit
P0071	Ambient Air Temperature Sensor Range/Performance
P0072	Ambient Air Temperature Sensor Circuit Low
P0073	Ambient Air Temperature Sensor Circuit High
P0074	Ambient Air Temperature Sensor Circuit Intermittent
P0075	Intake Valve Control Solenoid Circuit
P0076	Intake Valve Control Solenoid Circuit Low
P0077	Intake Valve Control Solenoid Circuit High
P0078	Exhaust Valve Control Solenoid Circuit
P0079	Exhaust Valve Control Solenoid Circuit Low
P0080	Exhaust Valve Control Solenoid Circuit High
P0081	Intake Valve Control Solenoid Circuit
P0082	Intake Valve Control Solenoid Circuit Low
P0083	Intake Valve Control Solenoid Circuit High
P0084	Exhaust Valve Control Solenoid Circuit
P0085	Exhaust Valve Control Solenoid Circuit Low
P0086	Exhaust Valve Control Solenoid Circuit High
P0087	Fuel Rail/System Pressure - Too Low
P0088	Fuel Rail/System Pressure - Too High
P0089	Fuel Pressure Regulator 1 Performance
P0090	Fuel Pressure Regulator 1 Control Circuit
P0091	Fuel Pressure Regulator 1 Control Circuit Low
P0092	Fuel Pressure Regulator 1 Control Circuit High
P0093	Fuel System Leak Detected - Large Leak
P0094	Fuel System Leak Detected - Small Leak
P0095	Intake Air Temperature Sensor 2 Circuit
P0096	Intake Air Temperature Sensor 2 Circuit Range/Performance
P0097	Intake Air Temperature Sensor 2 Circuit Low
P0098	Intake Air Temperature Sensor 2 Circuit High
P0099	Intake Air Temperature Sensor 2 Circuit Intermittent/Erratic

P0100 to P0199: Fuel Metering and Air Metering

Code	Description
P0100	Mass or Volume Air Flow Circuit
P0101	Mass or Volume Air Flow Circuit Range/Performance
P0102	Mass or Volume Air Flow Circuit Low Input
P0103	Mass or Volume Air Flow Circuit High Input
P0104	Mass or Volume Air Flow Circuit Intermittent
P0105	Manifold Absolute Pressure/Barometric Pressure Circuit
P0106	Manifold Absolute Pressure/Barometric Pressure Circuit Range/Performance
P0107	Manifold Absolute Pressure/Barometric Pressure Circuit Low Input
P0108	Manifold Absolute Pressure/Barometric Pressure Circuit High Input
P0109	Manifold Absolute Pressure/Barometric Pressure Circuit Intermittent
P0110	Intake Air Temperature Sensor 1 Circuit
P0111	Intake Air Temperature Sensor 1 Circuit Range/Performance
P0112	Intake Air Temperature Sensor 1 Circuit Low
P0113	Intake Air Temperature Sensor 1 Circuit High
P0114	Intake Air Temperature Sensor 1 Circuit Intermittent
P0115	Engine Coolant Temperature Circuit
P0116	Engine Coolant Temperature Circuit Range/Performance
P0117	Engine Coolant Temperature Circuit Low
P0118	Engine Coolant Temperature Circuit High
P0119	Engine Coolant Temperature Circuit Intermittent
P0120	Throttle/Pedal Position Sensor/Switch "A" Circuit
P0121	Throttle/Pedal Position Sensor/Switch "A" Circuit Range/Performance
P0122	Throttle/Pedal Position Sensor/Switch "A" Circuit Low
P0123	Throttle/Pedal Position Sensor/Switch "A" Circuit High
P0124	Throttle/Pedal Position Sensor/Switch "A" Circuit Intermittent
P0125	Insufficient Coolant Temperature for Closed-Loop Fuel Control
P0126	Insufficient Coolant Temperature for Stable Operation
P0127	Intake Air Temperature Too High
P0128	Coolant Thermostat (coolant temperature below thermostat regulating temperature)
P0129	Barometric Pressure Too Low
P0130	O2 Sensor Circuit
P0131	O2 Sensor Circuit Low Voltage
P0132	O2 Sensor Circuit High Voltage
P0133	O2 Sensor Circuit Slow Response
P0134	O2 Sensor Circuit No Activity Detected
P0135	O2 Sensor Heater Circuit
P0136	O2 Sensor Circuit
P0137	O2 Sensor Circuit Low Voltage
P0138	O2 Sensor Circuit High Voltage
P0139	O2 Sensor Circuit Slow Response
P0140	O2 Sensor Circuit No Activity Detected
P0141	O2 Sensor Heater Circuit
P0142	O2 Sensor Circuit
P0143	O2 Sensor Circuit Low Voltage
P0144	O2 Sensor Circuit High Voltage
P0145	O2 Sensor Circuit Slow Response
P0146	O2 Sensor Circuit No Activity Detected
P0147	O2 Sensor Heater Circuit
P0148	Fuel Delivery Error
P0149	Fuel Timing Error
P0150	O2 Sensor Circuit
P0151	O2 Sensor Circuit Low Voltage
P0152	O2 Sensor Circuit High Voltage
P0153	O2 Sensor Circuit Slow Response
P0154	O2 Sensor Circuit No Activity Detected
P0155	O2 Sensor Heater Circuit
P0156	O2 Sensor Circuit
P0157	O2 Sensor Circuit Low Voltage
P0158	O2 Sensor Circuit High Voltage
P0159	O2 Sensor Circuit Slow Response
P0160	O2 Sensor Circuit No Activity Detected
P0161	O2 Sensor Heater Circuit
P0162	O2 Sensor Circuit
P0163	O2 Sensor Circuit Low Voltage
P0164	O2 Sensor Circuit High Voltage

GENERIC OBD-II DTC CODES

P0165	O2 Sensor Circuit Slow Response
P0166	O2 Sensor Circuit No Activity Detected
P0167	O2 Sensor Heater Circuit
P0168	Fuel Temperature Too High
P0169	Incorrect Fuel Composition
P0170	Fuel Trim
P0171	System Too Lean
P0172	System Too Rich
P0173	Fuel Trim
P0174	System Too Lean
P0175	System Too Rich
P0176	Fuel Composition Sensor Circuit
P0177	Fuel Composition Sensor Circuit Range/Performance
P0178	Fuel Composition Sensor Circuit Low
P0179	Fuel Composition Sensor Circuit High
P0180	Fuel Temperature Sensor A Circuit
P0181	Fuel Temperature Sensor A Circuit Range/Performance
P0182	Fuel Temperature Sensor A Circuit Low
P0183	Fuel Temperature Sensor A Circuit High
P0184	Fuel Temperature Sensor A Circuit Intermittent
P0185	Fuel Temperature Sensor B Circuit
P0186	Fuel Temperature Sensor B Circuit Range/Performance
P0187	Fuel Temperature Sensor B Circuit Low
P0188	Fuel Temperature Sensor B Circuit High
P0189	Fuel Temperature Sensor B Circuit Intermittent
P0190	Fuel Rail Pressure Sensor Circuit
P0191	Fuel Rail Pressure Sensor Circuit Range/Performance
P0192	Fuel Rail Pressure Sensor Circuit Low
P0193	Fuel Rail Pressure Sensor Circuit High
P0194	Fuel Rail Pressure Sensor Circuit Intermittent
P0195	Engine Oil Temperature Sensor
P0196	Engine Oil Temperature Sensor Range/Performance
P0197	Engine Oil Temperature Sensor Low
P0198	Engine Oil Temperature Sensor High
P0199	Engine Oil Temperature Sensor Intermittent

P0200 to P0299: Fuel Metering, Air Metering and Injector Circuits

P0200	Injector Circuit/Open
P0201	Injector Circuit/Open - Cylinder 1
P0202	Injector Circuit/Open - Cylinder 2
P0203	Injector Circuit/Open - Cylinder 3
P0204	Injector Circuit/Open - Cylinder 4
P0205	Injector Circuit/Open - Cylinder 5
P0206	Injector Circuit/Open - Cylinder 6
P0207	Injector Circuit/Open - Cylinder 7
P0208	Injector Circuit/Open - Cylinder 8
P0209	Injector Circuit/Open - Cylinder 9
P0210	Injector Circuit/Open - Cylinder 10
P0211	Injector Circuit/Open - Cylinder 11
P0212	Injector Circuit/Open - Cylinder 12
P0213	Cold Start Injector 1
P0214	Cold Start Injector 2
P0215	Engine Shutoff Solenoid
P0216	Injector/Injection Timing Control Circuit
P0217	Engine Coolant Over Temperature Condition
P0218	Transmission Fluid Over Temperature Condition
P0219	Engine Overspeed Condition
P0220	Throttle/Pedal Position Sensor/Switch "B" Circuit
P0221	Throttle/Pedal Position Sensor/Switch "B" Circuit Range/Performance
P0222	Throttle/Pedal Position Sensor/Switch "B" Circuit Low
P0223	Throttle/Pedal Position Sensor/Switch "B" Circuit High
P0224	Throttle/Pedal Position Sensor/Switch "B" Circuit Intermittent
P0225	Throttle/Pedal Position Sensor/Switch "C" Circuit
P0226	Throttle/Pedal Position Sensor/Switch "C" Circuit Range/Performance
P0227	Throttle/Pedal Position Sensor/Switch "C" Circuit Low
P0228	Throttle/Pedal Position Sensor/Switch "C" Circuit High
P0229	Throttle/Pedal Position Sensor/Switch "C" Circuit Intermittent
P0230	Fuel Pump Primary Circuit
P0231	Fuel Pump Secondary Circuit Low
P0232	Fuel Pump Secondary Circuit High
P0233	Fuel Pump Secondary Circuit Intermittent
P0234	Turbo/Super Charger Overboost Condition
P0235	Turbo/Super Charger Boost Sensor "A" Circuit
P0236	Turbo/Super Charger Boost Sensor "A" Circuit Range/Performance
P0237	Turbo/Super Charger Boost Sensor "A" Circuit Low
P0238	Turbo/Super Charger Boost Sensor "A" Circuit High
P0239	Turbo/Super Charger Boost Sensor "B" Circuit
P0240	Turbo/Super Charger Boost Sensor "B" Circuit Range/Performance
P0241	Turbo/Super Charger Boost Sensor "B" Circuit Low
P0242	Turbo/Super Charger Boost Sensor "B" Circuit High
P0243	Turbo/Super Charger Wastegate Solenoid "A"
P0244	Turbo/Super Charger Wastegate Solenoid "A" Range/Performance
P0245	Turbo/Super Charger Wastegate Solenoid "A" Low
P0246	Turbo/Super Charger Wastegate Solenoid "A" High
P0247	Turbo/Super Charger Wastegate Solenoid "B"
P0248	Turbo/Super Charger Wastegate Solenoid "B" Range/Performance
P0249	Turbo/Super Charger Wastegate Solenoid "B" Low
P0250	Turbo/Super Charger Wastegate Solenoid "B" High
P0251	Injection Pump Fuel Metering Control "A" (Cam/Rotor/Injector)
P0252	Injection Pump Fuel Metering Control "A" Range/Performance (cam/rotor/injector)
P0253	Injection Pump Fuel Metering Control "A" Low (cam/rotor/injector)
P0254	Injection Pump Fuel Metering Control "A" High (cam/rotor/injector)
P0255	Injection Pump Fuel Metering Control "A" Intermittent (cam/rotor/injector)
P0256	Injection Pump Fuel Metering Control "B" (cam/rotor/injector)
P0257	Injection Pump Fuel Metering Control "B" Range/Performance (cam/rotor/injector)
P0258	Injection Pump Fuel Metering Control "B" Low (cam/rotor/injector)

APPENDIX B

P0259	Injection Pump Fuel Metering Control "B" High (cam/rotor/injector)
P0260	Injection Pump Fuel Metering Control "B" Intermittent (cam/rotor/injector)
P0261	Cylinder 1 Injector Circuit Low
P0262	Cylinder 1 Injector Circuit High
P0263	Cylinder 1 Contribution/Balance
P0264	Cylinder 2 Injector Circuit Low
P0265	Cylinder 2 Injector Circuit High
P0266	Cylinder 2 Contribution/Balance
P0267	Cylinder 3 Injector Circuit Low
P0268	Cylinder 3 Injector Circuit High
P0269	Cylinder 3 Contribution/Balance
P0270	Cylinder 4 Injector Circuit Low
P0271	Cylinder 4 Injector Circuit High
P0272	Cylinder 4 Contribution/Balance
P0273	Cylinder 5 Injector Circuit Low
P0274	Cylinder 5 Injector Circuit High
P0275	Cylinder 5 Contribution/Balance
P0276	Cylinder 6 Injector Circuit Low
P0277	Cylinder 6 Injector Circuit High
P0278	Cylinder 6 Contribution/Balance
P0279	Cylinder 7 Injector Circuit Low
P0280	Cylinder 7 Injector Circuit High
P0281	Cylinder 7 Contribution/Balance
P0282	Cylinder 8 Injector Circuit Low
P0283	Cylinder 8 Injector Circuit High
P0284	Cylinder 8 Contribution/Balance
P0285	Cylinder 9 Injector Circuit Low
P0286	Cylinder 9 Injector Circuit High
P0287	Cylinder 9 Contribution/Balance
P0288	Cylinder 10 Injector Circuit Low
P0289	Cylinder 10 Injector Circuit High
P0290	Cylinder 10 Contribution/Balance
P0291	Cylinder 11 Injector Circuit Low
P0292	Cylinder 11 Injector Circuit High
P0293	Cylinder 11 Contribution/Balance
P0294	Cylinder 12 Injector Circuit Low
P0295	Cylinder 12 Injector Circuit High
P0296	Cylinder 12 Contribution/Balance
P0297	Vehicle Overspeed Condition
P0298	Engine Oil Over Temperature
P0299	Turbo/Super Charger Underboost

P0300 to P0399: Ignition Systems and Misfires

P0300	Random/Multiple Cylinder Misfire Detected
P0301	Cylinder 1 Misfire Detected
P0302	Cylinder 2 Misfire Detected
P0303	Cylinder 3 Misfire Detected
P0304	Cylinder 4 Misfire Detected
P0305	Cylinder 5 Misfire Detected
P0306	Cylinder 6 Misfire Detected
P0307	Cylinder 7 Misfire Detected
P0308	Cylinder 8 Misfire Detected
P0309	Cylinder 9 Misfire Detected
P0310	Cylinder 10 Misfire Detected
P0311	Cylinder 11 Misfire Detected
P0312	Cylinder 12 Misfire Detected
P0313	Misfire Detected with Low Fuel
P0314	Single Cylinder Misfire (cylinder not specified)
P0315	Crankshaft Position System Variation Not Learned
P0316	Engine Misfire Detected on Startup (first 1,000 revolutions)
P0317	Rough Road Hardware Not Present
P0318	Rough Road Sensor "A" Signal Circuit
P0319	Rough Road Sensor "B"
P0320	Ignition/Distributor Engine Speed Input Circuit
P0321	Ignition/Distributor Engine Speed Input Circuit Range/Performance
P0322	Ignition/Distributor Engine Speed Input Circuit No Signal
P0323	Ignition/Distributor Engine Speed Input Circuit Intermittent
P0324	Knock Control System Error
P0325	Knock Sensor 1 Circuit
P0326	Knock Sensor 1 Circuit Range/Performance
P0327	Knock Sensor 1 Circuit Low
P0328	Knock Sensor 1 Circuit High
P0329	Knock Sensor 1 Circuit Input Intermittent
P0330	Knock Sensor 2 Circuit
P0331	Knock Sensor 2 Circuit Range/Performance
P0332	Knock Sensor 2 Circuit Low
P0333	Knock Sensor 2 Circuit High
P0334	Knock Sensor 2 Circuit Input Intermittent
P0335	Crankshaft Position Sensor "A" Circuit
P0336	Crankshaft Position Sensor "A" Circuit Range/Performance
P0337	Crankshaft Position Sensor "A" Circuit Low
P0338	Crankshaft Position Sensor "A" Circuit High
P0339	Crankshaft Position Sensor "A" Circuit Intermittent
P0340	Camshaft Position Sensor "A" Circuit
P0341	Camshaft Position Sensor "A" Circuit Range/Performance
P0342	Camshaft Position Sensor "A" Circuit Low
P0343	Camshaft Position Sensor "A" Circuit High
P0344	Camshaft Position Sensor "A" Circuit Intermittent
P0345	Camshaft Position Sensor "A" Circuit
P0346	Camshaft Position Sensor "A" Circuit Range/Performance
P0347	Camshaft Position Sensor "A" Circuit Low
P0348	Camshaft Position Sensor "A" Circuit High
P0349	Camshaft Position Sensor "A" Circuit Intermittent
P0350	Ignition Coil Primary/Secondary Circuit
P0351	Ignition Coil "A" Primary/Secondary Circuit
P0352	Ignition Coil "B" Primary/Secondary Circuit
P0353	Ignition Coil "C" Primary/Secondary Circuit
P0354	Ignition Coil "D" Primary/Secondary Circuit
P0355	Ignition Coil "E" Primary/Secondary Circuit
P0356	Ignition Coil "F" Primary/Secondary Circuit
P0357	Ignition Coil "G" Primary/Secondary Circuit
P0358	Ignition Coil "H" Primary/Secondary Circuit
P0359	Ignition Coil "I" Primary/Secondary Circuit
P0360	Ignition Coil "J" Primary/Secondary Circuit
P0361	Ignition Coil "K" Primary/Secondary Circuit
P0362	Ignition Coil "L" Primary/Secondary Circuit
P0363	Misfire Detected - Fueling Disabled

GENERIC OBD-II DTC CODES

P0364 Reserved
P0365 Camshaft Position Sensor "B" Circuit
P0366 Camshaft Position Sensor "B" Circuit Range/Performance
P0367 Camshaft Position Sensor "B" Circuit Low
P0368 Camshaft Position Sensor "B" Circuit High
P0369 Camshaft Position Sensor "B" Circuit Intermittent
P0370 Timing Reference High Resolution Signal "A"
P0371 Timing Reference High Resolution Signal "A" Too Many Pulses
P0372 Timing Reference High Resolution Signal "A" Too Few Pulses
P0373 Timing Reference High Resolution Signal "A" Intermittent/Erratic Pulses
P0374 Timing Reference High Resolution Signal "A" No Pulse
P0375 Timing Reference High Resolution Signal "B"
P0376 Timing Reference High Resolution Signal "B" Too Many Pulses
P0377 Timing Reference High Resolution Signal "B" Too Few Pulses
P0378 Timing Reference High Resolution Signal "B" Intermittent/Erratic Pulses
P0379 Timing Reference High Resolution Signal "B" No Pulses
P0380 Glow Plug/Heater Circuit "A"
P0381 Glow Plug/Heater Indicator Circuit
P0382 Glow Plug/Heater Circuit "B"
P0383 Reserved by SAE J2012
P0384 Reserved by SAE J2012
P0385 Crankshaft Position Sensor "B" Circuit
P0386 Crankshaft Position Sensor "B" Circuit Range/Performance
P0387 Crankshaft Position Sensor "B" Circuit Low
P0388 Crankshaft Position Sensor "B" Circuit High
P0389 Crankshaft Position Sensor "B" Circuit Intermittent
P0390 Camshaft Position Sensor "B" Circuit
P0391 Camshaft Position Sensor "B" Circuit Range/Performance
P0392 Camshaft Position Sensor "B" Circuit Low
P0393 Camshaft Position Sensor "B" Circuit High
P0394 Camshaft Position Sensor "B" Circuit Intermittent

P0400 to P0499: Auxiliary Emission Controls

P0400 Exhaust Gas Recirculation Flow
P0401 Exhaust Gas Recirculation Flow Insufficient Detected
P0402 Exhaust Gas Recirculation Flow Excessive Detected
P0403 Exhaust Gas Recirculation Control Circuit
P0404 Exhaust Gas Recirculation Control Circuit Range/Performance
P0405 Exhaust Gas Recirculation Sensor "A" Circuit Low
P0406 Exhaust Gas Recirculation Sensor "A" Circuit High
P0407 Exhaust Gas Recirculation Sensor "B" Circuit Low
P0408 Exhaust Gas Recirculation Sensor "B" Circuit High
P0409 Exhaust Gas Recirculation Sensor "A" Circuit
P0410 Secondary Air Injection System
P0411 Secondary Air Injection System Incorrect Flow Detected
P0412 Secondary Air Injection System Switching Valve "A" Circuit
P0413 Secondary Air Injection System Switching Valve "A" Circuit Open
P0414 Secondary Air Injection System Switching Valve "A" Circuit Shorted
P0415 Secondary Air Injection System Switching Valve "B" Circuit
P0416 Secondary Air Injection System Switching Valve "B" Circuit Open
P0417 Secondary Air Injection System Switching Valve "B" Circuit Shorted
P0418 Secondary Air Injection System Control "A" Circuit
P0419 Secondary Air Injection System Control "B" Circuit
P0420 Catalyst System Efficiency Below Threshold
P0421 Warm Up Catalyst Efficiency Below Threshold
P0422 Main Catalyst Efficiency Below Threshold
P0423 Heated Catalyst Efficiency Below Threshold
P0424 Heated Catalyst Temperature Below Threshold
P0425 Catalyst Temperature Sensor
P0426 Catalyst Temperature Sensor Range/Performance
P0427 Catalyst Temperature Sensor Low
P0428 Catalyst Temperature Sensor High
P0429 Catalyst Heater Control Circuit
P0430 Catalyst System Efficiency Below Threshold
P0431 Warm Up Catalyst Efficiency Below Threshold
P0432 Main Catalyst Efficiency Below Threshold
P0433 Heated Catalyst Efficiency Below Threshold
P0434 Heated Catalyst Temperature Below Threshold
P0435 Catalyst Temperature Sensor
P0436 Catalyst Temperature Sensor Range/Performance
P0437 Catalyst Temperature Sensor Low
P0438 Catalyst Temperature Sensor High
P0439 Catalyst Heater Control Circuit
P0440 Evaporative Emission System
P0441 Evaporative Emission System Incorrect Purge Flow
P0442 Evaporative Emission System Leak Detected (small leak)
P0443 Evaporative Emission System Purge Control Valve Circuit
P0444 Evaporative Emission System Purge Control Valve Circuit Open
P0445 Evaporative Emission System Purge Control Valve Circuit Shorted
P0446 Evaporative Emission System Vent Control Circuit
P0447 Evaporative Emission System Vent Control Circuit Open
P0448 Evaporative Emission System Vent Control Circuit Shorted
P0449 Evaporative Emission System Vent Valve/Solenoid Circuit
P0450 Evaporative Emission System Pressure Sensor/Switch
P0451 Evaporative Emission System Pressure Sensor/Switch Range/Performance
P0452 Evaporative Emission System Pressure Sensor/Switch Low
P0453 Evaporative Emission System Pressure Sensor/Switch High
P0454 Evaporative Emission System Pressure Sensor/Switch Intermittent
P0455 Evaporative Emission System Leak Detected (large leak)
P0456 Evaporative Emission System Leak Detected (very small leak)
P0457 Evaporative Emission System Leak Detected (fuel cap loose/off)
P0458 Evaporative Emission System Purge Control Valve Circuit Low
P0459 Evaporative Emission System Purge Control Valve Circuit High
P0460 Fuel Level Sensor "A" Circuit
P0461 Fuel Level Sensor "A" Circuit Range/Performance
P0462 Fuel Level Sensor "A" Circuit Low
P0463 Fuel Level Sensor "A" Circuit High
P0464 Fuel Level Sensor "A" Circuit Intermittent
P0465 EVAP Purge Flow Sensor Circuit

APPENDIX B

P0466 EVAP Purge Flow Sensor Circuit Range/Performance
P0467 EVAP Purge Flow Sensor Circuit Low
P0468 EVAP Purge Flow Sensor Circuit High
P0469 EVAP Purge Flow Sensor Circuit Intermittent
P0470 Exhaust Pressure Sensor
P0471 Exhaust Pressure Sensor Range/Performance
P0472 Exhaust Pressure Sensor Low
P0473 Exhaust Pressure Sensor High
P0474 Exhaust Pressure Sensor Intermittent
P0475 Exhaust Pressure Control Valve
P0476 Exhaust Pressure Control Valve Range/Performance
P0477 Exhaust Pressure Control Valve Low
P0478 Exhaust Pressure Control Valve High
P0479 Exhaust Pressure Control Valve Intermittent
P0480 Fan 1 Control Circuit
P0481 Fan 2 Control Circuit
P0482 Fan 3 Control Circuit
P0483 Fan Rationality Check
P0484 Fan Circuit Over Current
P0485 Fan Power/Ground Circuit
P0486 Exhaust Gas Recirculation Sensor "B" Circuit
P0487 Exhaust Gas Recirculation Throttle Position Control Circuit
P0488 Exhaust Gas Recirculation Throttle Position Control Range/Performance
P0489 Exhaust Gas Recirculation Control Circuit Low
P0490 Exhaust Gas Recirculation Control Circuit High
P0491 Secondary Air Injection System Insufficient Flow
P0492 Secondary Air Injection System Insufficient Flow
P0493 Fan Overspeed
P0494 Fan Speed Low
P0495 Fan Speed High
P0496 Evaporative Emission System High Purge Flow
P0497 Evaporative Emission System Low Purge Flow
P0498 Evaporative Emission System Vent Valve Control Circuit Low
P0499 Evaporative Emission System Vent Valve Control Circuit High

P0500 to P0599: Vehicle Speed Controls and Idle System Controls
P0500 Vehicle Speed Sensor "A"
P0501 Vehicle Speed Sensor "A" Range/Performance
P0502 Vehicle Speed Sensor "A" Circuit Low Input
P0503 Vehicle Speed Sensor "A" Intermittent/Erratic/High
P0504 Brake Switch "A"/"B" Correlation
P0505 Idle Air Control System
P0506 Idle Air Control System RPM Lower Than Expected
P0507 Idle Air Control System RPM Higher Than Expected
P0508 Idle Air Control System Circuit Low
P0509 Idle Air Control System Circuit High
P0510 Closed Throttle Position Switch
P0511 Idle Air Control Circuit
P0512 Starter Request Circuit
P0513 Incorrect Immobilizer Key
P0514 Battery Temperature Sensor Circuit Range/Performance
P0515 Battery Temperature Sensor Circuit
P0516 Battery Temperature Sensor Circuit Low
P0517 Battery Temperature Sensor Circuit High
P0518 Idle Air Control Circuit Intermittent
P0519 Idle Air Control System Performance
P0520 Engine Oil Pressure Sensor/Switch Circuit
P0521 Engine Oil Pressure Sensor/Switch Range/Performance
P0522 Engine Oil Pressure Sensor/Switch Low Voltage
P0523 Engine Oil Pressure Sensor/Switch High Voltage
P0524 Engine Oil Pressure Too Low
P0525 Cruise Control Servo Control Circuit Range/Performance
P0526 Fan Speed Sensor Circuit
P0527 Fan Speed Sensor Circuit Range/Performance
P0528 Fan Speed Sensor Circuit No Signal
P0529 Fan Speed Sensor Circuit Intermittent
P0530 A/C Refrigerant Pressure Sensor "A" Circuit
P0531 A/C Refrigerant Pressure Sensor "A" Circuit Range/Performance
P0532 A/C Refrigerant Pressure Sensor "A" Circuit Low
P0533 A/C Refrigerant Pressure Sensor "A" Circuit High
P0534 A/C Refrigerant Charge Loss
P0535 A/C Evaporator Temperature Sensor Circuit
P0536 A/C Evaporator Temperature Sensor Circuit Range/Performance
P0537 A/C Evaporator Temperature Sensor Circuit Low
P0538 A/C Evaporator Temperature Sensor Circuit High
P0539 A/C Evaporator Temperature Sensor Circuit Intermittent
P0540 Intake Air Heater "A" Circuit
P0541 Intake Air Heater "A" Circuit Low
P0542 Intake Air Heater "A" Circuit High
P0543 Intake Air Heater "A" Circuit Open
P0544 Exhaust Gas Temperature Sensor Circuit
P0545 Exhaust Gas Temperature Sensor Circuit Low
P0546 Exhaust Gas Temperature Sensor Circuit High
P0547 Exhaust Gas Temperature Sensor Circuit
P0548 Exhaust Gas Temperature Sensor Circuit Low
P0549 Exhaust Gas Temperature Sensor Circuit High
P0550 Power Steering Pressure Sensor/Switch Circuit
P0551 Power Steering Pressure Sensor/Switch Circuit Range/Performance
P0552 Power Steering Pressure Sensor/Switch Circuit Low Input
P0553 Power Steering Pressure Sensor/Switch Circuit High Input
P0554 Power Steering Pressure Sensor/Switch Circuit Intermittent
P0555 Brake Booster Pressure Sensor Circuit
P0556 Brake Booster Pressure Sensor Circuit Range/Performance
P0557 Brake Booster Pressure Sensor Circuit Low Input
P0558 Brake Booster Pressure Sensor Circuit High Input
P0559 Brake Booster Pressure Sensor Circuit Intermittent
P0560 System Voltage
P0561 System Voltage Unstable
P0562 System Voltage Low
P0563 System Voltage High
P0564 Cruise Control Multi-Function Input "A" Circuit
P0565 Cruise Control On Signal
P0566 Cruise Control Off Signal
P0567 Cruise Control Resume Signal
P0568 Cruise Control Set Signal
P0569 Cruise Control Coast Signal
P0570 Cruise Control Accelerate Signal
P0571 Brake Switch "A" Circuit

GENERIC OBD-II DTC CODES

P0572 Brake Switch "A" Circuit Low
P0573 Brake Switch "A" Circuit High
P0574 Cruise Control System - Vehicle Speed Too High
P0575 Cruise Control Input Circuit
P0576 Cruise Control Input Circuit Low
P0577 Cruise Control Input Circuit High
P0578 Cruise Control Multi-Function Input "A" Circuit Stuck
P0579 Cruise Control Multi-Function Input "A" Circuit Range/Performance
P0580 Cruise Control Multi-Function Input "A" Circuit Low
P0581 Cruise Control Multi-Function Input "A" Circuit High
P0582 Cruise Control Vacuum Control Circuit/Open
P0583 Cruise Control Vacuum Control Circuit Low
P0584 Cruise Control Vacuum Control Circuit High
P0585 Cruise Control Multi-Function Input "A"/"B" Correlation
P0586 Cruise Control Vent Control Circuit/Open
P0587 Cruise Control Vent Control Circuit Low
P0588 Cruise Control Vent Control Circuit High
P0589 Cruise Control Multi-Function Input "B" Circuit
P0590 Cruise Control Multi-Function Input "B" Circuit Stuck
P0591 Cruise Control Multi-Function Input "B" Circuit Range/Performance
P0592 Cruise Control Multi-Function Input "B" Circuit Low
P0593 Cruise Control Multi-Function Input "B" Circuit High
P0594 Cruise Control Servo Control Circuit/Open
P0595 Cruise Control Servo Control Circuit Low
P0596 Cruise Control Servo Control Circuit High
P0597 Thermostat Heater Control Circuit/Open
P0598 Thermostat Heater Control Circuit Low
P0599 Thermostat Heater Control Circuit High

P0600 to P0699: Computer Open Circuit

P0600 Serial Communication Link
P0601 Internal Control Module Memory Check Sum Error
P0602 Control Module Programming Error
P0603 Internal Control Module Keep Alive Memory (KAM) Error
P0604 Internal Control Module Random Access Memory (RAM) Error
P0605 Internal Control Module Read Only Memory (ROM) Error
P0606 ECM/PCM Processor
P0607 Control Module Performance
P0608 Control Module VSS Output "A"
P0609 Control Module VSS Output "B"
P0610 Control Module Vehicle Options Error
P0611 Fuel Injector Control Module Performance
P0612 Fuel Injector Control Module Relay Control
P0613 TCM Processor
P0614 ECM/TCM Incompatible
P0615 Starter Relay Circuit
P0616 Starter Relay Circuit Low
P0617 Starter Relay Circuit High
P0618 Alternative Fuel Control Module KAM Error
P0619 Alternative Fuel Control Module RAM/ROM Error
P0620 Generator Control Circuit
P0621 Generator Lamp/L Terminal Circuit
P0622 Generator Field/F Terminal Circuit
P0623 Generator Lamp Control Circuit
P0624 Fuel Cap Lamp Control Circuit
P0625 Generator Field/F Terminal Circuit Low
P0626 Generator Field/F Terminal Circuit High
P0627 Fuel Pump "A" Control Circuit /Open
P0628 Fuel Pump "A" Control Circuit Low
P0629 Fuel Pump "A" Control Circuit High
P0630 VIN Not Programmed or Incompatible - ECM/PCM
P0631 VIN Not Programmed or Incompatible - TCM
P0632 Odometer Not Programmed - ECM/PCM
P0633 Immobilizer Key Not Programmed - ECM/PCM
P0634 PCM/ECM/TCM Internal Temperature Too High
P0635 Power Steering Control Circuit
P0636 Power Steering Control Circuit Low
P0637 Power Steering Control Circuit High
P0638 Throttle Actuator Control Range/Performance
P0639 Throttle Actuator Control Range/Performance
P0640 Intake Air Heater Control Circuit
P0641 Sensor Reference Voltage "A" Circuit/Open
P0642 Sensor Reference Voltage "A" Circuit Low
P0643 Sensor Reference Voltage "A" Circuit High
P0644 Driver Display Serial Communication Circuit
P0645 A/C Clutch Relay Control Circuit
P0646 A/C Clutch Relay Control Circuit Low
P0647 A/C Clutch Relay Control Circuit High
P0648 Immobilizer Lamp Control Circuit
P0649 Speed Control Lamp Control Circuit
P0650 Malfunction Indicator Lamp (MIL) Control Circuit
P0651 Sensor Reference Voltage "B" Circuit/Open
P0652 Sensor Reference Voltage "B" Circuit Low
P0653 Sensor Reference Voltage "B" Circuit High
P0654 Engine RPM Output Circuit
P0655 Engine Hot Lamp Output Control Circuit
P0656 Fuel Level Output Circuit
P0657 Actuator Supply Voltage "A" Circuit/Open
P0658 Actuator Supply Voltage "A" Circuit Low
P0659 Actuator Supply Voltage "A" Circuit High
P0660 Intake Manifold Tuning Valve Control Circuit/Open
P0661 Intake Manifold Tuning Valve Control Circuit Low
P0662 Intake Manifold Tuning Valve Control Circuit High
P0663 Intake Manifold Tuning Valve Control Circuit/Open
P0664 Intake Manifold Tuning Valve Control Circuit Low
P0665 Intake Manifold Tuning Valve Control Circuit High
P0666 PCM/ECM/TCM Internal Temperature Sensor Circuit
P0667 PCM/ECM/TCM Internal Temperature Sensor Range/Performance
P0668 PCM/ECM/TCM Internal Temperature Sensor Circuit Low
P0669 PCM/ECM/TCM Internal Temperature Sensor Circuit High
P0670 Glow Plug Module Control Circuit
P0671 Cylinder 1 Glow Plug Circuit
P0672 Cylinder 2 Glow Plug Circuit
P0673 Cylinder 3 Glow Plug Circuit
P0674 Cylinder 4 Glow Plug Circuit
P0675 Cylinder 5 Glow Plug Circuit

P0676 Cylinder 6 Glow Plug Circuit
P0677 Cylinder 7 Glow Plug Circuit
P0678 Cylinder 8 Glow Plug Circuit
P0679 Cylinder 9 Glow Plug Circuit
P0680 Cylinder 10 Glow Plug Circuit
P0681 Cylinder 11 Glow Plug Circuit
P0682 Cylinder 12 Glow Plug Circuit
P0683 Glow Plug Control Module to PCM Communication Circuit
P0684 Glow Plug Control Module to PCM Communication Circuit Range/Performance
P0685 ECM/PCM Power Relay Control Circuit/Open
P0686 ECM/PCM Power Relay Control Circuit Low
P0687 ECM/PCM Power Relay Control Circuit High
P0688 ECM/PCM Power Relay Sense Circuit/Open
P0689 ECM/PCM Power Relay Sense Circuit Low
P0690 ECM/PCM Power Relay Sense Circuit High
P0691 Fan 1 Control Circuit Low
P0692 Fan 1 Control Circuit High
P0693 Fan 2 Control Circuit Low
P0694 Fan 2 Control Circuit High
P0695 Fan 3 Control Circuit Low
P0696 Fan 3 Control Circuit High
P0697 Sensor Reference Voltage "C" Circuit/Open
P0698 Sensor Reference Voltage "C" Circuit Low
P0699 Sensor Reference Voltage "C" Circuit High

P0700 to P1000: Transmission Controls

P0700 Transmission Control System (MIL Request)
P0701 Transmission Control System Range/Performance
P0702 Transmission Control System Electrical
P0703 Brake Switch "B" Circuit
P0704 Clutch Switch Input Circuit Malfunction
P0705 Transmission Range Sensor Circuit Malfunction (PRNDL input)
P0706 Transmission Range Sensor Circuit Range/Performance
P0707 Transmission Range Sensor Circuit Low
P0708 Transmission Range Sensor Circuit High
P0709 Transmission Range Sensor Circuit Intermittent
P0710 Transmission Fluid Temperature Sensor "A" Circuit
P0711 Transmission Fluid Temperature Sensor "A" Circuit Range/Performance
P0712 Transmission Fluid Temperature Sensor "A" Circuit Low
P0713 Transmission Fluid Temperature Sensor "A" Circuit High
P0714 Transmission Fluid Temperature Sensor "A" Circuit Intermittent
P0715 Input/Turbine Speed Sensor "A" Circuit
P0716 Input/Turbine Speed Sensor "A" Circuit Range/Performance
P0717 Input/Turbine Speed Sensor "A" Circuit No Signal
P0718 Input/Turbine Speed Sensor "A" Circuit Intermittent
P0719 Brake Switch "B" Circuit Low
P0720 Output Speed Sensor Circuit
P0721 Output Speed Sensor Circuit Range/Performance
P0722 Output Speed Sensor Circuit No Signal
P0723 Output Speed Sensor Circuit Intermittent
P0724 Brake Switch "B" Circuit High
P0725 Engine Speed Input Circuit
P0726 Engine Speed Input Circuit Range/Performance
P0727 Engine Speed Input Circuit No Signal
P0728 Engine Speed Input Circuit Intermittent
P0729 Gear 6 Incorrect Ratio
P0730 Incorrect Gear Ratio
P0731 Gear 1 Incorrect Ratio
P0732 Gear 2 Incorrect Ratio
P0733 Gear 3 Incorrect Ratio
P0734 Gear 4 Incorrect Ratio
P0735 Gear 5 Incorrect Ratio
P0736 Reverse Incorrect Ratio
P0737 TCM Engine Speed Output Circuit
P0738 TCM Engine Speed Output Circuit Low
P0739 TCM Engine Speed Output Circuit High
P0740 Torque Converter Clutch Circuit/Open
P0741 Torque Converter Clutch Circuit Performance or Stuck Off
P0742 Torque Converter Clutch Circuit Stuck On
P0743 Torque Converter Clutch Circuit Electrical
P0744 Torque Converter Clutch Circuit Intermittent
P0745 Pressure Control Solenoid "A"
P0746 Pressure Control Solenoid "A" Performance or Stuck Off
P0747 Pressure Control Solenoid "A" Stuck On
P0748 Pressure Control Solenoid "A" Electrical
P0749 Pressure Control Solenoid "A" Intermittent
P0750 Shift Solenoid "A"
P0751 Shift Solenoid "A" Performance or Stuck Off
P0752 Shift Solenoid "A" Stuck On
P0753 Shift Solenoid "A" Electrical
P0754 Shift Solenoid "A" Intermittent
P0755 Shift Solenoid "B"
P0756 Shift Solenoid "B" Performance or Stuck Off
P0757 Shift Solenoid "B" Stuck On
P0758 Shift Solenoid "B" Electrical
P0759 Shift Solenoid "B" Intermittent
P0760 Shift Solenoid "C"
P0761 Shift Solenoid "C" Performance or Stuck Off
P0762 Shift Solenoid "C" Stuck On
P0763 Shift Solenoid "C" Electrical
P0764 Shift Solenoid "C" Intermittent
P0765 Shift Solenoid "D"
P0766 Shift Solenoid "D" Performance or Stuck Off
P0767 Shift Solenoid "D" Stuck On
P0768 Shift Solenoid "D" Electrical
P0769 Shift Solenoid "D" Intermittent
P0770 Shift Solenoid "E"
P0771 Shift Solenoid "E" Performance or Stuck Off
P0772 Shift Solenoid "E" Stuck On
P0773 Shift Solenoid "E" Electrical
P0774 Shift Solenoid "E" Intermittent
P0775 Pressure Control Solenoid "B"
P0776 Pressure Control Solenoid "B" Performance or Stuck Off
P0777 Pressure Control Solenoid "B" Stuck On
P0778 Pressure Control Solenoid "B" Electrical
P0779 Pressure Control Solenoid "B" Intermittent
P0780 Shift Error

GENERIC OBD-II DTC CODES

Code	Description
P0781	1-2 Shift
P0782	2-3 Shift
P0783	3-4 Shift
P0784	4-5 Shift
P0785	Shift/Timing Solenoid
P0786	Shift/Timing Solenoid Range/Performance
P0787	Shift/Timing Solenoid Low
P0788	Shift/Timing Solenoid High
P0789	Shift/Timing Solenoid Intermittent
P0790	Normal/Performance Switch Circuit
P0791	Intermediate Shaft Speed Sensor "A" Circuit
P0792	Intermediate Shaft Speed Sensor "A" Circuit Range/Performance
P0793	Intermediate Shaft Speed Sensor "A" Circuit No Signal
P0794	Intermediate Shaft Speed Sensor "A" Circuit Intermittent
P0795	Pressure Control Solenoid "C"
P0796	Pressure Control Solenoid "C" Performance or Stuck Off
P0797	Pressure Control Solenoid "C" Stuck On
P0798	Pressure Control Solenoid "C" Electrical
P0799	Pressure Control Solenoid "C" Intermittent
P0800	Transfer Case Control System (MIL request)
P0801	Reverse Inhibit Control Circuit
P0802	Transmission Control System MIL Request Circuit/Open
P0803	1-4 Upshift (Skip Shift) Solenoid Control Circuit
P0804	1-4 Upshift (Skip Shift) Lamp Control Circuit
P0805	Clutch Position Sensor Circuit
P0806	Clutch Position Sensor Circuit Range/Performance
P0807	Clutch Position Sensor Circuit Low
P0808	Clutch Position Sensor Circuit High
P0809	Clutch Position Sensor Circuit Intermittent
P0810	Clutch Position Control Error
P0811	Excessive Clutch Slippage
P0812	Reverse Input Circuit
P0813	Reverse Output Circuit
P0814	Transmission Range Display Circuit
P0815	Upshift Switch Circuit
P0816	Downshift Switch Circuit
P0817	Starter Disable Circuit
P0818	Driveline Disconnect Switch Input Circuit
P0819	Up and Down Shift Switch to Transmission Range Correlation
P0820	Gear Lever X-Y Position Sensor Circuit
P0821	Gear Lever X Position Circuit
P0822	Gear Lever Y Position Circuit
P0823	Gear Lever X Position Circuit Intermittent
P0824	Gear Lever Y Position Circuit Intermittent
P0825	Gear Lever Push-Pull Switch (shift anticipate)
P0826	Up and Down Shift Switch Circuit
P0827	Up and Down Shift Switch Circuit Low
P0828	Up and Down Shift Switch Circuit High
P0829	5-6 Shift
P0830	Clutch Pedal Switch "A" Circuit
P0831	Clutch Pedal Switch "A" Circuit Low
P0832	Clutch Pedal Switch "A" Circuit High
P0833	Clutch Pedal Switch "B" Circuit
P0834	Clutch Pedal Switch "B" Circuit Low
P0835	Clutch Pedal Switch "B" Circuit High
P0836	Four Wheel Drive (4WD) Switch Circuit
P0837	Four Wheel Drive (4WD) Switch Circuit Range/Performance
P0838	Four Wheel Drive (4WD) Switch Circuit Low
P0839	Four Wheel Drive (4WD) Switch Circuit High
P0840	Transmission Fluid Pressure Sensor/Switch "A" Circuit
P0841	Transmission Fluid Pressure Sensor/Switch "A" Circuit Range/Performance
P0842	Transmission Fluid Pressure Sensor/Switch "A" Circuit Low
P0843	Transmission Fluid Pressure Sensor/Switch "A" Circuit High
P0844	Transmission Fluid Pressure Sensor/Switch "A" Circuit Intermittent
P0845	Transmission Fluid Pressure Sensor/Switch "B" Circuit
P0846	Transmission Fluid Pressure Sensor/Switch "B" Circuit Range/Performance
P0847	Transmission Fluid Pressure Sensor/Switch "B" Circuit Low
P0848	Transmission Fluid Pressure Sensor/Switch "B" Circuit High
P0849	Transmission Fluid Pressure Sensor/Switch "B" Circuit Intermittent
P0850	Park/Neutral Switch Input Circuit
P0851	Park/Neutral Switch Input Circuit Low
P0852	Park/Neutral Switch Input Circuit High
P0853	Drive Switch Input Circuit
P0854	Drive Switch Input Circuit Low
P0855	Drive Switch Input Circuit High
P0856	Traction Control Input Signal
P0857	Traction Control Input Signal Range/Performance
P0858	Traction Control Input Signal Low
P0859	Traction Control Input Signal High
P0860	Gear Shift Module Communication Circuit
P0861	Gear Shift Module Communication Circuit Low
P0862	Gear Shift Module Communication Circuit High
P0863	TCM Communication Circuit
P0864	TCM Communication Circuit Range/Performance
P0865	TCM Communication Circuit Low
P0866	TCM Communication Circuit High
P0867	Transmission Fluid Pressure
P0868	Transmission Fluid Pressure Low
P0869	Transmission Fluid Pressure High
P0870	Transmission Fluid Pressure Sensor/Switch "C" Circuit
P0871	Transmission Fluid Pressure Sensor/Switch "C" Circuit Range/Performance
P0872	Transmission Fluid Pressure Sensor/Switch "C" Circuit Low
P0873	Transmission Fluid Pressure Sensor/Switch "C" Circuit High
P0874	Transmission Fluid Pressure Sensor/Switch "C" Circuit Intermittent
P0875	Transmission Fluid Pressure Sensor/Switch "D" Circuit
P0876	Transmission Fluid Pressure Sensor/Switch "D" Circuit Range/Performance
P0877	Transmission Fluid Pressure Sensor/Switch "D" Circuit Low
P0878	Transmission Fluid Pressure Sensor/Switch "D" Circuit High
P0879	Transmission Fluid Pressure Sensor/Switch "D" Circuit Intermittent
P0880	TCM Power Input Signal

APPENDIX B

P0881	TCM Power Input Signal Range/Performance	P0936	Hydraulic Pressure Sensor Circuit Intermittent
P0882	TCM Power Input Signal Low	P0937	Hydraulic Oil Temperature Sensor Circuit
P0883	TCM Power Input Signal High	P0938	Hydraulic Oil Temperature Sensor Range/Performance
P0884	TCM Power Input Signal Intermittent	P0939	Hydraulic Oil Temperature Sensor Circuit Low
P0885	TCM Power Relay Control Circuit/Open	P0940	Hydraulic Oil Temperature Sensor Circuit High
P0886	TCM Power Relay Control Circuit Low	P0941	Hydraulic Oil Temperature Sensor Circuit Intermittent
P0887	TCM Power Relay Control Circuit High	P0942	Hydraulic Pressure Unit
P0888	TCM Power Relay Sense Circuit	P0943	Hydraulic Pressure Unit Cycling Period Too Short
P0889	TCM Power Relay Sense Circuit Range/Performance	P0944	Hydraulic Pressure Unit Loss of Pressure
P0890	TCM Power Relay Sense Circuit Low	P0945	Hydraulic Pump Relay Circuit/Open
P0891	TCM Power Relay Sense Circuit High	P0946	Hydraulic Pump Relay Circuit Range/Performance
P0892	TCM Power Relay Sense Circuit Intermittent	P0947	Hydraulic Pump Relay Circuit Low
P0893	Multiple Gears Engaged	P0948	Hydraulic Pump Relay Circuit High
P0894	Transmission Component Slipping	P0949	Auto Shift Manual Adaptive Learning Not Complete
P0895	Shift Time Too Short	P0950	Auto Shift Manual Control Circuit
P0896	Shift Time Too Long	P0951	Auto Shift Manual Control Circuit Range/Performance
P0897	Transmission Fluid Deteriorated	P0952	Auto Shift Manual Control Circuit Low
P0898	Transmission Control System MIL Request Circuit Low	P0953	Auto Shift Manual Control Circuit High
P0899	Transmission Control System MIL Request Circuit High	P0954	Auto Shift Manual Control Circuit Intermittent
P0900	Clutch Actuator Circuit/Open	P0955	Auto Shift Manual Mode Circuit
P0901	Clutch Actuator Circuit Range/Performance	P0956	Auto Shift Manual Mode Circuit Range/Performance
P0902	Clutch Actuator Circuit Low	P0957	Auto Shift Manual Mode Circuit Low
P0903	Clutch Actuator Circuit High	P0958	Auto Shift Manual Mode Circuit High
P0904	Gate Select Position Circuit	P0959	Auto Shift Manual Mode Circuit Intermittent
P0905	Gate Select Position Circuit Range/Performance	P0960	Pressure Control Solenoid "A" Control Circuit/Open
P0906	Gate Select Position Circuit Low	P0961	Pressure Control Solenoid "A" Control Circuit Range/Performance
P0907	Gate Select Position Circuit High		
P0908	Gate Select Position Circuit Intermittent	P0962	Pressure Control Solenoid "A" Control Circuit Low
P0909	Gate Select Control Error	P0963	Pressure Control Solenoid "A" Control Circuit High
P0910	Gate Select Actuator Circuit/Open	P0964	Pressure Control Solenoid "B" Control Circuit/Open
P0911	Gate Select Actuator Circuit Range/Performance	P0965	Pressure Control Solenoid "B" Control Circuit Range/Performance
P0912	Gate Select Actuator Circuit Low		
P0913	Gate Select Actuator Circuit High	P0966	Pressure Control Solenoid "B" Control Circuit Low
P0914	Gear Shift Position Circuit	P0967	Pressure Control Solenoid "B" Control Circuit High
P0915	Gear Shift Position Circuit Range/Performance	P0968	Pressure Control Solenoid "C" Control Circuit/Open
P0916	Gear Shift Position Circuit Low	P0969	Pressure Control Solenoid "C" Control Circuit Range/Performance
P0917	Gear Shift Position Circuit High		
P0918	Gear Shift Position Circuit Intermittent	P0970	Pressure Control Solenoid "C" Control Circuit Low
P0919	Gear Shift Position Control Error	P0971	Pressure Control Solenoid "C" Control Circuit High
P0920	Gear Shift Forward Actuator Circuit/Open	P0972	Shift Solenoid "A" Control Circuit Range/Performance
P0921	Gear Shift Forward Actuator Circuit Range/Performance	P0973	Shift Solenoid "A" Control Circuit Low
P0922	Gear Shift Forward Actuator Circuit Low	P0974	Shift Solenoid "A" Control Circuit High
P0923	Gear Shift Forward Actuator Circuit High	P0975	Shift Solenoid "B" Control Circuit Range/Performance
P0924	Gear Shift Reverse Actuator Circuit/Open	P0976	Shift Solenoid "B" Control Circuit Low
P0925	Gear Shift Reverse Actuator Circuit Range/Performance	P0977	Shift Solenoid "B" Control Circuit High
P0926	Gear Shift Reverse Actuator Circuit Low	P0978	Shift Solenoid "C" Control Circuit Range/Performance
P0927	Gear Shift Reverse Actuator Circuit High	P0979	Shift Solenoid "C" Control Circuit Low
P0928	Gear Shift Lock Solenoid Control Circuit/Open	P0980	Shift Solenoid "C" Control Circuit High
P0929	Gear Shift Lock Solenoid Control Circuit Range/Performance	P0981	Shift Solenoid "D" Control Circuit Range/Performance
P0930	Gear Shift Lock Solenoid Control Circuit Low	P0982	Shift Solenoid "D" Control Circuit Low
P0931	Gear Shift Lock Solenoid Control Circuit High	P0983	Shift Solenoid "D" Control Circuit High
P0932	Hydraulic Pressure Sensor Circuit	P0984	Shift Solenoid "E" Control Circuit Range/Performance
P0933	Hydraulic Pressure Sensor Range/Performance	P0985	Shift Solenoid "E" Control Circuit Low
P0934	Hydraulic Pressure Sensor Circuit Low	P0986	Shift Solenoid "E" Control Circuit High
P0935	Hydraulic Pressure Sensor Circuit High	P0987	Transmission Fluid Pressure Sensor/Switch "E" Circuit

GENERIC OBD-II DTC CODES

Code	Description
P0988	Transmission Fluid Pressure Sensor/Switch "E" Circuit Range/Performance
P0989	Transmission Fluid Pressure Sensor/Switch "E" Circuit Low
P0990	Transmission Fluid Pressure Sensor/Switch "E" Circuit High
P0991	Transmission Fluid Pressure Sensor/Switch "E" Circuit Intermittent
P0992	Transmission Fluid Pressure Sensor/Switch "F" Circuit
P0993	Transmission Fluid Pressure Sensor/Switch "F" Circuit Range/Performance
P0994	Transmission Fluid Pressure Sensor/Switch "F" Circuit Low
P0995	Transmission Fluid Pressure Sensor/Switch "F" Circuit High
P0996	Transmission Fluid Pressure Sensor/Switch "F" Circuit Intermittent
P0997	Shift Solenoid "F" Control Circuit Range/Performance
P0998	Shift Solenoid "F" Control Circuit Low
P0999	Shift Solenoid "F" Control Circuit High

P0A00 to P0A99: System Sensors and Voltage Controls

Code	Description
P0A00	Motor Electronics Coolant Temperature Sensor Circuit
P0A01	Motor Electronics Coolant Temperature Sensor Circuit Range/Performance
P0A02	Motor Electronics Coolant Temperature Sensor Circuit Low
P0A03	Motor Electronics Coolant Temperature Sensor Circuit High
P0A04	Motor Electronics Coolant Temperature Sensor Circuit Intermittent
P0A05	Motor Electronics Coolant Pump Control Circuit/Open
P0A06	Motor Electronics Coolant Pump Control Circuit Low
P0A07	Motor Electronics Coolant Pump Control Circuit High
P0A08	DC/DC Converter Status Circuit
P0A09	DC/DC Converter Status Circuit Low Input
P0A10	DC/DC Converter Status Circuit High Input
P0A11	DC/DC Converter Enable Circuit/Open
P0A12	DC/DC Converter Enable Circuit Low
P0A13	DC/DC Converter Enable Circuit High
P0A14	Engine Mount Control Circuit/Open
P0A15	Engine Mount Control Circuit Low
P0A16	Engine Mount Control Circuit High
P0A17	Motor Torque Sensor Circuit
P0A18	Motor Torque Sensor Circuit Range/Performance
P0A19	Motor Torque Sensor Circuit Low
P0A20	Motor Torque Sensor Circuit High
P0A21	Motor Torque Sensor Circuit Intermittent
P0A22	Generator Torque Sensor Circuit
P0A23	Generator Torque Sensor Circuit Range/Performance
P0A24	Generator Torque Sensor Circuit Low
P0A25	Generator Torque Sensor Circuit High
P0A26	Generator Torque Sensor Circuit Intermittent
P0A27	Battery Power Off Circuit
P0A28	Battery Power Off Circuit Low
P0A29	Battery Power Off Circuit High

P2000 to P2FFF: Intake Controls, Throttle Body Controls and Auxiliary Emission Controls

Code	Description
P2000	NOx Trap Efficiency Below Threshold
P2001	NOx Trap Efficiency Below Threshold
P2002	Particulate Trap Efficiency Below Threshold
P2003	Particulate Trap Efficiency Below Threshold
P2004	Intake Manifold Runner Control Stuck Open
P2005	Intake Manifold Runner Control Stuck Open
P2006	Intake Manifold Runner Control Stuck Closed
P2007	Intake Manifold Runner Control Stuck Closed
P2008	Intake Manifold Runner Control Circuit/Open
P2009	Intake Manifold Runner Control Circuit Low
P2010	Intake Manifold Runner Control Circuit High
P2011	Intake Manifold Runner Control Circuit/Open
P2012	Intake Manifold Runner Control Circuit Low
P2013	Intake Manifold Runner Control Circuit High
P2014	Intake Manifold Runner Position Sensor/Switch Circuit
P2015	Intake Manifold Runner Position Sensor/Switch Circuit Range/Performance
P2016	Intake Manifold Runner Position Sensor/Switch Circuit Low
P2017	Intake Manifold Runner Position Sensor/Switch Circuit High
P2018	Intake Manifold Runner Position Sensor/Switch Circuit Intermittent
P2019	Intake Manifold Runner Position Sensor/Switch Circuit
P2020	Intake Manifold Runner Position Sensor/Switch Circuit Range/Performance
P2021	Intake Manifold Runner Position Sensor/Switch Circuit Low
P2022	Intake Manifold Runner Position Sensor/Switch Circuit High
P2023	Intake Manifold Runner Position Sensor/Switch Circuit Intermittent
P2024	Evaporative Emissions (EVAP) Fuel Vapor Temperature Sensor Circuit
P2025	Evaporative Emissions (EVAP) Fuel Vapor Temperature Sensor Performance
P2026	Evaporative Emissions (EVAP) Fuel Vapor Temperature Sensor Circuit Low Voltage
P2027	Evaporative Emissions (EVAP) Fuel Vapor Temperature Sensor Circuit High Voltage
P2028	Evaporative Emissions (EVAP) Fuel Vapor Temperature Sensor Circuit Intermittent
P2029	Fuel Fired Heater Disabled
P2030	Fuel Fired Heater Performance
P2031	Exhaust Gas Temperature Sensor Circuit
P2032	Exhaust Gas Temperature Sensor Circuit Low
P2033	Exhaust Gas Temperature Sensor Circuit High
P2034	Exhaust Gas Temperature Sensor Circuit
P2035	Exhaust Gas Temperature Sensor Circuit Low
P2036	Exhaust Gas Temperature Sensor Circuit High
P2037	Reductant Injection Air Pressure Sensor Circuit
P2038	Reductant Injection Air Pressure Sensor Circuit Range/Performance
P2039	Reductant Injection Air Pressure Sensor Circuit Low Input
P2040	Reductant Injection Air Pressure Sensor Circuit High Input
P2041	Reductant Injection Air Pressure Sensor Circuit Intermittent
P2042	Reductant Temperature Sensor Circuit
P2043	Reductant Temperature Sensor Circuit Range/Performance
P2044	Reductant Temperature Sensor Circuit Low Input
P2045	Reductant Temperature Sensor Circuit High Input
P2046	Reductant Temperature Sensor Circuit Intermittent

APPENDIX B

P2047	Reductant Injector Circuit/Open	P2100	Throttle Actuator Control Motor Circuit/Open
P2048	Reductant Injector Circuit Low	P2101	Throttle Actuator Control Motor Circuit Range/Performance
P2049	Reductant Injector Circuit High	P2102	Throttle Actuator Control Motor Circuit Low
P2050	Reductant Injector Circuit/Open	P2103	Throttle Actuator Control Motor Circuit High
P2051	Reductant Injector Circuit Low	P2104	Throttle Actuator Control System - Forced Idle
P2052	Reductant Injector Circuit High	P2105	Throttle Actuator Control System - Forced Engine Shutdown
P2053	Reductant Injector Circuit/Open	P2106	Throttle Actuator Control System - Forced Limited Power
P2054	Reductant Injector Circuit Low	P2107	Throttle Actuator Control Module Processor
P2055	Reductant Injector Circuit High	P2108	Throttle Actuator Control Module Performance
P2056	Reductant Injector Circuit/Open	P2109	Throttle/Pedal Position Sensor "A" Minimum Stop Performance
P2057	Reductant Injector Circuit Low	P2110	Throttle Actuator Control System - Forced Limited RPM
P2058	Reductant Injector Circuit High	P2111	Throttle Actuator Control System - Stuck Open
P2059	Reductant Injection Air Pump Control Circuit/Open	P2112	Throttle Actuator Control System - Stuck Closed
P2060	Reductant Injection Air Pump Control Circuit Low	P2113	Throttle/Pedal Position Sensor "B" Minimum Stop Performance
P2061	Reductant Injection Air Pump Control Circuit High	P2114	Throttle/Pedal Position Sensor "C" Minimum Stop Performance
P2062	Reductant Supply Control Circuit/Open	P2115	Throttle/Pedal Position Sensor "D" Minimum Stop Performance
P2063	Reductant Supply Control Circuit Low	P2116	Throttle/Pedal Position Sensor "E" Minimum Stop Performance
P2064	Reductant Supply Control Circuit High	P2117	Throttle/Pedal Position Sensor "F" Minimum Stop Performance
P2065	Fuel Level Sensor "B" Circuit	P2118	Throttle Actuator Control Motor Current Range/Performance
P2066	Fuel Level Sensor "B" Performance	P2119	Throttle Actuator Control Throttle Body Range/Performance
P2067	Fuel Level Sensor "B" Circuit Low	P2120	Throttle/Pedal Position Sensor/Switch "D" Circuit
P2068	Fuel Level Sensor "B" Circuit High	P2121	Throttle/Pedal Position Sensor/Switch "D" Circuit Range/Performance
P2069	Fuel Level Sensor "B" Circuit Intermittent	P2122	Throttle/Pedal Position Sensor/Switch "D" Circuit Low Input
P2070	Intake Manifold Tuning (IMT) Valve Stuck Open	P2123	Throttle/Pedal Position Sensor/Switch "D" Circuit High Input
P2071	Intake Manifold Tuning (IMT) Valve Stuck Closed	P2124	Throttle/Pedal Position Sensor/Switch "D" Circuit Intermittent
P2075	Intake Manifold Tuning (IMT) Valve Position Sensor/Switch Circuit	P2125	Throttle/Pedal Position Sensor/Switch "E" Circuit
P2076	Intake Manifold Tuning (IMT) Valve Position Sensor/Switch Circuit Range/Performance	P2126	Throttle/Pedal Position Sensor/Switch "E" Circuit Range/Performance
P2077	Intake Manifold Tuning (IMT) Valve Position Sensor/Switch Circuit Low	P2127	Throttle/Pedal Position Sensor/Switch "E" Circuit Low Input
P2078	Intake Manifold Tuning (IMT) Valve Position Sensor/Switch Circuit High	P2128	Throttle/Pedal Position Sensor/Switch "E" Circuit High Input
P2079	Intake Manifold Tuning (IMT) Valve Position Sensor/Switch Circuit Intermittent	P2129	Throttle/Pedal Position Sensor/Switch "E" Circuit Intermittent
P2080	Exhaust Gas Temperature Sensor Circuit Range/Performance	P2130	Throttle/Pedal Position Sensor/Switch "F" Circuit
P2081	Exhaust Gas Temperature Sensor Circuit Intermittent	P2131	Throttle/Pedal Position Sensor/Switch "F" Circuit Range Performance
P2082	Exhaust Gas Temperature Sensor Circuit Range/Performance	P2132	Throttle/Pedal Position Sensor/Switch "F" Circuit Low Input
P2083	Exhaust Gas Temperature Sensor Circuit Intermittent	P2133	Throttle/Pedal Position Sensor/Switch "F" Circuit High Input
P2084	Exhaust Gas Temperature Sensor Circuit Range/Performance	P2134	Throttle/Pedal Position Sensor/Switch "F" Circuit Intermittent
P2085	Exhaust Gas Temperature Sensor Circuit Intermittent	P2135	Throttle/Pedal Position Sensor/Switch "A"/"B" Voltage Correlation
P2086	Exhaust Gas Temperature Sensor Circuit Range/Performance	P2136	Throttle/Pedal Position Sensor/Switch "A"/"C" Voltage Correlation
P2087	Exhaust Gas Temperature Sensor Circuit Intermittent	P2137	Throttle/Pedal Position Sensor/Switch "B"/"C" Voltage Correlation
P2088	"A" Camshaft Position Actuator Control Circuit Low		
P2089	"A" Camshaft Position Actuator Control Circuit High		
P2090	"B" Camshaft Position Actuator Control Circuit Low		
P2091	"B" Camshaft Position Actuator Control Circuit High		
P2092	"A" Camshaft Position Actuator Control Circuit Low	P2138	Throttle/Pedal Position Sensor/Switch "D"/"E" Voltage Correlation
P2093	"A" Camshaft Position Actuator Control Circuit High		
P2094	"B" Camshaft Position Actuator Control Circuit Low	P2139	Throttle/Pedal Position Sensor/Switch "D"/"F" Voltage Correlation
P2095	"B" Camshaft Position Actuator Control Circuit High		
P2096	Post Catalyst Fuel Trim System Too Lean		
P2097	Post Catalyst Fuel Trim System Too Rich		
P2098	Post Catalyst Fuel Trim System Too Lean		
P2099	Post Catalyst Fuel Trim System Too Rich		

GENERIC OBD-II DTC CODES

Code	Description
P2140	Throttle/Pedal Position Sensor/Switch "E"/"F" Voltage Correlation
P2141	Exhaust Gas Recirculation Throttle Control Circuit Low
P2142	Exhaust Gas Recirculation Throttle Control Circuit High
P2143	Exhaust Gas Recirculation Vent Control Circuit/Open
P2144	Exhaust Gas Recirculation Vent Control Circuit Low
P2145	Exhaust Gas Recirculation Vent Control Circuit High
P2146	Fuel Injector Group "A" Supply Voltage Circuit/Open
P2147	Fuel Injector Group "A" Supply Voltage Circuit Low
P2148	Fuel Injector Group "A" Supply Voltage Circuit High
P2149	Fuel Injector Group "B" Supply Voltage Circuit/Open
P2150	Fuel Injector Group "B" Supply Voltage Circuit Low
P2151	Fuel Injector Group "B" Supply Voltage Circuit High
P2152	Fuel Injector Group "C" Supply Voltage Circuit/Open
P2153	Fuel Injector Group "C" Supply Voltage Circuit Low
P2154	Fuel Injector Group "C" Supply Voltage Circuit High
P2155	Fuel Injector Group "D" Supply Voltage Circuit/Open
P2156	Fuel Injector Group "D" Supply Voltage Circuit Low
P2157	Fuel Injector Group "D" Supply Voltage Circuit High
P2158	Vehicle Speed Sensor "B"
P2159	Vehicle Speed Sensor "B" Range/Performance
P2160	Vehicle Speed Sensor "B" Circuit Low
P2161	Vehicle Speed Sensor "B" Intermittent/Erratic
P2162	Vehicle Speed Sensor "A"/"B" Correlation
P2163	Throttle/Pedal Position Sensor "A" Maximum Stop Performance
P2164	Throttle/Pedal Position Sensor "B" Maximum Stop Performance
P2165	Throttle/Pedal Position Sensor "C" Maximum Stop Performance
P2166	Throttle/Pedal Position Sensor "D" Maximum Stop Performance
P2167	Throttle/Pedal Position Sensor "E" Maximum Stop Performance
P2168	Throttle/Pedal Position Sensor "F" Maximum Stop Performance
P2169	Exhaust Pressure Regulator Vent Solenoid Control Circuit/Open
P2170	Exhaust Pressure Regulator Vent Solenoid Control Circuit Low
P2171	Exhaust Pressure Regulator Vent Solenoid Control Circuit High
P2172	Throttle Actuator Control System - Sudden High Airflow Detected
P2173	Throttle Actuator Control System - High Airflow Detected
P2174	Throttle Actuator Control System - Sudden Low Airflow Detected
P2175	Throttle Actuator Control System - Low Airflow Detected
P2176	Throttle Actuator Control System - Idle Position Not Learned
P2177	System Too Lean Off Idle
P2178	System Too Rich Off Idle
P2179	System Too Lean Off Idle
P2180	System Too Rich Off Idle
P2181	Cooling System Performance
P2182	Engine Coolant Temperature Sensor 2 Circuit
P2183	Engine Coolant Temperature Sensor 2 Circuit Range/Performance
P2184	Engine Coolant Temperature Sensor 2 Circuit Low
P2185	Engine Coolant Temperature Sensor 2 Circuit High
P2186	Engine Coolant Temperature Sensor 2 Circuit Intermittent/Erratic
P2187	System Too Lean at Idle
P2188	System Too Rich at Idle
P2189	System Too Lean at Idle
P2190	System Too Rich at Idle
P2191	System Too Lean at Higher Load
P2192	System Too Rich at Higher Load
P2193	System Too Lean at Higher Load
P2194	System Too Rich at Higher Load
P2195	O2 Sensor Signal Stuck Lean
P2196	O2 Sensor Signal Stuck Rich
P2197	O2 Sensor Signal Stuck Lean
P2198	O2 Sensor Signal Stuck Rich
P2199	Intake Air Temperature Sensor 1/2 Correlation
P2200	NOx Sensor Circuit
P2201	NOx Sensor Circuit Range/Performance
P2202	NOx Sensor Circuit Low Input
P2203	NOx Sensor Circuit High Input
P2204	NOx Sensor Circuit Intermittent Input
P2205	NOx Sensor Heater Control Circuit/Open
P2206	NOx Sensor Heater Control Circuit Low
P2207	NOx Sensor Heater Control Circuit High
P2208	NOx Sensor Heater Sense Circuit
P2209	NOx Sensor Heater Sense Circuit Range/Performance
P2210	NOx Sensor Heater Sense Circuit Low Input
P2211	NOx Sensor Heater Sense Circuit High Input
P2212	NOx Sensor Heater Sense Circuit Intermittent
P2213	NOx Sensor Circuit
P2214	NOx Sensor Circuit Range/Performance
P2215	NOx Sensor Circuit Low Input
P2216	NOx Sensor Circuit High Input
P2217	NOx Sensor Circuit Intermittent Input
P2218	NOx Sensor Heater Control Circuit/Open
P2219	NOx Sensor Heater Control Circuit Low
P2220	NOx Sensor Heater Control Circuit High
P2221	NOx Sensor Heater Sense Circuit
P2222	NOx Sensor Heater Sense Circuit Range/Performance
P2223	NOx Sensor Heater Sense Circuit Low
P2224	NOx Sensor Heater Sense Circuit High
P2225	NOx Sensor Heater Sense Circuit Intermittent
P2226	Barometric Pressure Circuit
P2227	Barometric Pressure Circuit Range/Performance
P2228	Barometric Pressure Circuit Low
P2229	Barometric Pressure Circuit High
P2230	Barometric Pressure Circuit Intermittent
P2231	O2 Sensor Signal Circuit Shorted to Heater Circuit
P2232	O2 Sensor Signal Circuit Shorted to Heater Circuit
P2233	O2 Sensor Signal Circuit Shorted to Heater Circuit
P2234	O2 Sensor Signal Circuit Shorted to Heater Circuit
P2235	O2 Sensor Signal Circuit Shorted to Heater Circuit
P2236	O2 Sensor Signal Circuit Shorted to Heater Circuit
P2237	O2 Sensor Positive Current Control Circuit/Open
P2238	O2 Sensor Positive Current Control Circuit Low
P2239	O2 Sensor Positive Current Control Circuit High
P2240	O2 Sensor Positive Current Control Circuit/Open
P2241	O2 Sensor Positive Current Control Circuit Low
P2242	O2 Sensor Positive Current Control Circuit High
P2243	O2 Sensor Reference Voltage Circuit/Open
P2244	O2 Sensor Reference Voltage Performance
P2245	O2 Sensor Reference Voltage Circuit Low

APPENDIX B

Code	Description
P2246	O2 Sensor Reference Voltage Circuit High
P2247	O2 Sensor Reference Voltage Circuit/Open
P2248	O2 Sensor Reference Voltage Performance
P2249	O2 Sensor Reference Voltage Circuit Low
P2250	O2 Sensor Reference Voltage Circuit High
P2251	O2 Sensor Negative Current Control Circuit/Open
P2252	O2 Sensor Negative Current Control Circuit Low
P2253	O2 Sensor Negative Current Control Circuit High
P2254	O2 Sensor Negative Current Control Circuit/Open
P2255	O2 Sensor Negative Current Control Circuit Low
P2256	O2 Sensor Negative Current Control Circuit High
P2257	Secondary Air Injection System Control "A" Circuit Low
P2258	Secondary Air Injection System Control "A" Circuit High
P2259	Secondary Air Injection System Control "B" Circuit Low
P2260	Secondary Air Injection System Control "B" Circuit High
P2261	Turbo/Super Charger Bypass Valve - Mechanical
P2262	Turbo Boost Pressure Not Detected - Mechanical
P2263	Turbo/Super Charger Boost System Performance
P2264	Water in Fuel Sensor Circuit
P2265	Water in Fuel Sensor Circuit Range/Performance
P2266	Water in Fuel Sensor Circuit Low
P2267	Water in Fuel Sensor Circuit High
P2268	Water in Fuel Sensor Circuit Intermittent
P2269	Water in Fuel Condition
P2270	O2 Sensor Signal Stuck Lean
P2271	O2 Sensor Signal Stuck Rich
P2272	O2 Sensor Signal Stuck Lean
P2273	O2 Sensor Signal Stuck Rich
P2274	O2 Sensor Signal Stuck Lean
P2275	O2 Sensor Signal Stuck Rich
P2276	O2 Sensor Signal Stuck Lean
P2277	O2 Sensor Signal Stuck Rich
P2278	O2 Sensor Signals Swapped Bank 1 Sensor 3/Bank 2 Sensor 3
P2279	Intake Air System Leak
P2280	Air Flow Restriction/Air Leak Between Air Filter and MAF
P2281	Air Leak Between MAF and Throttle Body
P2282	Air Leak Between Throttle Body and Intake Valves
P2283	Injector Control Pressure Sensor Circuit
P2284	Injector Control Pressure Sensor Circuit Range/Performance
P2285	Injector Control Pressure Sensor Circuit Low
P2286	Injector Control Pressure Sensor Circuit High
P2287	Injector Control Pressure Sensor Circuit Intermittent
P2288	Injector Control Pressure Too High
P2289	Injector Control Pressure Too High - Engine Off
P2290	Injector Control Pressure Too Low
P2291	Injector Control Pressure Too Low - Engine Cranking
P2292	Injector Control Pressure Erratic
P2293	Fuel Pressure Regulator 2 Performance
P2294	Fuel Pressure Regulator 2 Control Circuit
P2295	Fuel Pressure Regulator 2 Control Circuit Low
P2296	Fuel Pressure Regulator 2 Control Circuit High
P2297	O2 Sensor Out of Range During Deceleration
P2298	O2 Sensor Out of Range During Deceleration
P2299	Brake Pedal Position/Accelerator Pedal Position Incompatible
P2300	Ignition Coil "A" Primary Control Circuit Low
P2301	Ignition Coil "A" Primary Control Circuit High
P2302	Ignition Coil "A" Secondary Circuit
P2303	Ignition Coil "B" Primary Control Circuit Low
P2304	Ignition Coil "B" Primary Control Circuit High
P2305	Ignition Coil "B" Secondary Circuit
P2306	Ignition Coil "C" Primary Control Circuit Low
P2307	Ignition Coil "C" Primary Control Circuit High
P2308	Ignition Coil "C" Secondary Circuit
P2309	Ignition Coil "D" Primary Control Circuit Low
P2310	Ignition Coil "D" Primary Control Circuit High
P2311	Ignition Coil "D" Secondary Circuit
P2312	Ignition Coil "E" Primary Control Circuit Low
P2313	Ignition Coil "E" Primary Control Circuit High
P2314	Ignition Coil "E" Secondary Circuit
P2315	Ignition Coil "F" Primary Control Circuit Low
P2316	Ignition Coil "F" Primary Control Circuit High
P2317	Ignition Coil "F" Secondary Circuit
P2318	Ignition Coil "G" Primary Control Circuit Low
P2319	Ignition Coil "G" Primary Control Circuit High
P2320	Ignition Coil "G" Secondary Circuit
P2321	Ignition Coil "H" Primary Control Circuit Low
P2322	Ignition Coil "H" Primary Control Circuit High
P2323	Ignition Coil "H" Secondary Circuit
P2324	Ignition Coil "I" Primary Control Circuit Low
P2325	Ignition Coil "I" Primary Control Circuit High
P2326	Ignition Coil "I" Secondary Circuit
P2327	Ignition Coil "J" Primary Control Circuit Low
P2328	Ignition Coil "J" Primary Control Circuit High
P2329	Ignition Coil "J" Secondary Circuit
P2330	Ignition Coil "K" Primary Control Circuit Low
P2331	Ignition Coil "K" Primary Control Circuit High
P2332	Ignition Coil "K" Secondary Circuit
P2333	Ignition Coil "L" Primary Control Circuit Low
P2334	Ignition Coil "L" Primary Control Circuit High
P2335	Ignition Coil "L" Secondary Circuit
P2336	Cylinder #1 Above Knock Threshold
P2337	Cylinder #2 Above Knock Threshold
P2338	Cylinder #3 Above Knock Threshold
P2339	Cylinder #4 Above Knock Threshold
P2340	Cylinder #5 Above Knock Threshold
P2341	Cylinder #6 Above Knock Threshold
P2342	Cylinder #7 Above Knock Threshold
P2343	Cylinder #8 Above Knock Threshold
P2344	Cylinder #9 Above Knock Threshold
P2345	Cylinder #10 Above Knock Threshold
P2346	Cylinder #11 Above Knock Threshold
P2347	Cylinder #12 Above Knock Threshold
P2400	Evaporative Emission System Leak Detection Pump Control Circuit/Open
P2401	Evaporative Emission System Leak Detection Pump Control Circuit Low
P2402	Evaporative Emission System Leak Detection Pump Control Circuit High
P2403	Evaporative Emission System Leak Detection Pump Sense Circuit/Open

GENERIC OBD-II DTC CODES

Code	Description
P2404	Evaporative Emission System Leak Detection Pump Sense Circuit Range/Performance
P2405	Evaporative Emission System Leak Detection Pump Sense Circuit Low
P2406	Evaporative Emission System Leak Detection Pump Sense Circuit High
P2407	Evaporative Emission System Leak Detection Pump Sense Circuit Intermittent/Erratic
P2408	Fuel Cap Sensor/Switch Circuit
P2409	Fuel Cap Sensor/Switch Circuit Range/Performance
P2410	Fuel Cap Sensor/Switch Circuit Low
P2411	Fuel Cap Sensor/Switch Circuit High
P2412	Fuel Cap Sensor/Switch Circuit Intermittent/Erratic
P2413	Exhaust Gas Recirculation System Performance
P2414	O2 Sensor Exhaust Sample Error
P2415	O2 Sensor Exhaust Sample Error
P2416	O2 Sensor Signals Swapped Bank 1 Sensor 2/Bank 1 Sensor 3
P2417	O2 Sensor Signals Swapped Bank 2 Sensor 2/Bank 2 Sensor 3
P2418	Evaporative Emission System Switching Valve Control Circuit/Open
P2419	Evaporative Emission System Switching Valve Control Circuit Low
P2420	Evaporative Emission System Switching Valve Control Circuit High
P2421	Evaporative Emission System Vent Valve Stuck Open
P2422	Evaporative Emission System Vent Valve Stuck Closed
P2423	HC Adsorption Catalyst Efficiency Below Threshold
P2424	HC Adsorption Catalyst Efficiency Below Threshold
P2425	Exhaust Gas Recirculation Cooling Valve Control Circuit/Open
P2426	Exhaust Gas Recirculation Cooling Valve Control Circuit Low
P2427	Exhaust Gas Recirculation Cooling Valve Control Circuit High
P2428	Exhaust Gas Temperature Too High
P2429	Exhaust Gas Temperature Too High
P2430	Secondary Air Injection System Airflow/Pressure Sensor Circuit
P2431	Secondary Air Injection System Airflow/Pressure Sensor Circuit Range/Performance
P2432	Secondary Air Injection System Airflow/Pressure Sensor Circuit Low
P2433	Secondary Air Injection System Airflow/Pressure Sensor Circuit High
P2434	Secondary Air Injection System Airflow/Pressure Sensor Circuit Intermittent/Erratic
P2435	Secondary Air Injection System Airflow/Pressure Sensor Circuit
P2436	Secondary Air Injection System Airflow/Pressure Sensor Circuit Range/Performance
P2437	Secondary Air Injection System Airflow/Pressure Sensor Circuit Low
P2438	Secondary Air Injection System Airflow/Pressure Sensor Circuit High
P2439	Secondary Air Injection System Airflow/Pressure Sensor Circuit Intermittent/Erratic
P2440	Secondary Air Injection System Switching Valve Stuck Open
P2441	Secondary Air Injection System Switching Valve Stuck Closed
P2442	Secondary Air Injection System Switching Valve Stuck Open
P2443	Secondary Air Injection System Switching Valve Stuck Closed
P2444	Secondary Air Injection System Pump Stuck On
P2445	Secondary Air Injection System Pump Stuck Off
P2446	Secondary Air Injection System Pump Stuck On
P2447	Secondary Air Injection System Pump Stuck Off
P2500	Generator Lamp/L-Terminal Circuit Low
P2501	Generator Lamp/L-Terminal Circuit High
P2502	Charging System Voltage
P2503	Charging System Voltage Low
P2504	Charging System Voltage High
P2505	ECM/PCM Power Input Signal
P2506	ECM/PCM Power Input Signal Range/Performance
P2507	ECM/PCM Power Input Signal Low
P2508	ECM/PCM Power Input Signal High
P2509	ECM/PCM Power Input Signal Intermittent
P2510	ECM/PCM Power Relay Sense Circuit Range/Performance
P2511	ECM/PCM Power Relay Sense Circuit Intermittent
P2512	Event Data Recorder Request Circuit/ Open
P2513	Event Data Recorder Request Circuit Low
P2514	Event Data Recorder Request Circuit High
P2515	A/C Refrigerant Pressure Sensor "B" Circuit
P2516	A/C Refrigerant Pressure Sensor "B" Circuit Range/Performance
P2517	A/C Refrigerant Pressure Sensor "B" Circuit Low
P2518	A/C Refrigerant Pressure Sensor "B" Circuit High
P2519	A/C Request "A" Circuit
P2520	A/C Request "A" Circuit Low
P2521	A/C Request "A" Circuit High
P2522	A/C Request "B" Circuit
P2523	A/C Request "B" Circuit Low
P2524	A/C Request "B" Circuit High
P2525	Vacuum Reservoir Pressure Sensor Circuit
P2526	Vacuum Reservoir Pressure Sensor Circuit Range/Performance
P2527	Vacuum Reservoir Pressure Sensor Circuit Low
P2528	Vacuum Reservoir Pressure Sensor Circuit High
P2529	Vacuum Reservoir Pressure Sensor Circuit Intermittent
P2530	Ignition Switch Run Position Circuit
P2531	Ignition Switch Run Position Circuit Low
P2532	Ignition Switch Run Position Circuit High
P2533	Ignition Switch Run/Start Position Circuit
P2534	Ignition Switch Run/Start Position Circuit Low
P2535	Ignition Switch Run/Start Position Circuit High
P2536	Ignition Switch Accessory Position Circuit
P2537	Ignition Switch Accessory Position Circuit Low
P2538	Ignition Switch Accessory Position Circuit High
P2539	Low Pressure Fuel System Sensor Circuit
P2540	Low Pressure Fuel System Sensor Circuit Range/Performance
P2541	Low Pressure Fuel System Sensor Circuit Low
P2542	Low Pressure Fuel System Sensor Circuit High
P2543	Low Pressure Fuel System Sensor Circuit Intermittent
P2544	Torque Management Request Input Signal "A"
P2545	Torque Management Request Input Signal "A" Range/Performance
P2546	Torque Management Request Input Signal "A" Low
P2547	Torque Management Request Input Signal "A" High

APPENDIX B

P2548	Torque Management Request Input Signal "B"	P2614	Camshaft Position Signal Output Circuit/Open
P2549	Torque Management Request Input Signal "B" Range/Performance	P2615	Camshaft Position Signal Output Circuit Low
P2550	Torque Management Request Input Signal "B" Low	P2616	Camshaft Position Signal Output Circuit High
P2551	Torque Management Request Input Signal "B" High	P2617	Crankshaft Position Signal Output Circuit/Open
P2552	Throttle/Fuel Inhibit Circuit	P2618	Crankshaft Position Signal Output Circuit Low
P2553	Throttle/Fuel Inhibit Circuit Range/Performance	P2619	Crankshaft Position Signal Output Circuit High
P2554	Throttle/Fuel Inhibit Circuit Low	P2620	Throttle Position Output Circuit/Open
P2555	Throttle/Fuel Inhibit Circuit High	P2621	Throttle Position Output Circuit Low
P2556	Engine Coolant Level Sensor/Switch Circuit	P2622	Throttle Position Output Circuit High
P2557	Engine Coolant Level Sensor/Switch Circuit Range/Performance	P2623	Injector Control Pressure Regulator Circuit/Open
P2558	Engine Coolant Level Sensor/Switch Circuit Low	P2624	Injector Control Pressure Regulator Circuit Low
P2559	Engine Coolant Level Sensor/Switch Circuit High	P2625	Injector Control Pressure Regulator Circuit High
P2560	Engine Coolant Level Low	P2626	O2 Sensor Pumping Current Trim Circuit/Open
P2561	A/C Control Module Requested MIL Illumination	P2627	O2 Sensor Pumping Current Trim Circuit Low
P2562	Turbocharger Boost Control Position Sensor Circuit	P2628	O2 Sensor Pumping Current Trim Circuit High
P2563	Turbocharger Boost Control Position Sensor Circuit Range/Performance	P2629	O2 Sensor Pumping Current Trim Circuit/Open
P2564	Turbocharger Boost Control Position Sensor Circuit Low	P2630	O2 Sensor Pumping Current Trim Circuit Low
P2565	Turbocharger Boost Control Position Sensor Circuit High	P2631	O2 Sensor Pumping Current Trim Circuit High
P2566	Turbocharger Boost Control Position Sensor Circuit Intermittent	P2632	Fuel Pump "B" Control Circuit /Open
P2567	Direct Ozone Reduction Catalyst Temperature Sensor Circuit	P2633	Fuel Pump "B" Control Circuit Low
P2568	Direct Ozone Reduction Catalyst Temperature Sensor Circuit Range/Performance	P2634	Fuel Pump "B" Control Circuit High
P2569	Direct Ozone Reduction Catalyst Temperature Sensor Circuit Low	P2635	Fuel Pump "A" Low Flow/Performance
P2570	Direct Ozone Reduction Catalyst Temperature Sensor Circuit High	P2636	Fuel Pump "B" Low Flow/Performance
P2571	Direct Ozone Reduction Catalyst Temperature Sensor Circuit Intermittent/Erratic	P2637	Torque Management Feedback Signal "A"
P2572	Direct Ozone Reduction Catalyst Deterioration Sensor Circuit	P2638	Torque Management Feedback Signal "A" Range/Performance
P2573	Direct Ozone Reduction Catalyst Deterioration Sensor Circuit Range/Performance	P2639	Torque Management Feedback Signal "A" Low
P2574	Direct Ozone Reduction Catalyst Deterioration Sensor Circuit Low	P2640	Torque Management Feedback Signal "A" High
P2575	Direct Ozone Reduction Catalyst Deterioration Sensor Circuit High	P2641	Torque Management Feedback Signal "B"
P2576	Direct Ozone Reduction Catalyst Deterioration Sensor Circuit Intermittent/Erratic	P2642	Torque Management Feedback Signal "B" Range/Performance
P2577	Direct Ozone Reduction Catalyst Efficiency Below Threshold	P2643	Torque Management Feedback Signal "B" Low
P2600	Coolant Pump Control Circuit/Open	P2644	Torque Management Feedback Signal "B" High
P2601	Coolant Pump Control Circuit Range/Performance	P2645	"A" Rocker Arm Actuator Control Circuit/Open
P2602	Coolant Pump Control Circuit Low	P2646	"A" Rocker Arm Actuator System Performance or Stuck Off
P2603	Coolant Pump Control Circuit High	P2647	"A" Rocker Arm Actuator System Stuck On
P2604	Intake Air Heater "A" Circuit Range/Performance	P2648	"A" Rocker Arm Actuator Control Circuit Low
P2605	Intake Air Heater "A" Circuit/Open	P2649	"A" Rocker Arm Actuator Control Circuit High
P2606	Intake Air Heater "B" Circuit Range/Performance	P2650	"B" Rocker Arm Actuator Control Circuit/Open
P2607	Intake Air Heater "B" Circuit Low	P2651	"B" Rocker Arm Actuator System Performance or Stuck Off
P2608	Intake Air Heater "B" Circuit High	P2652	"B" Rocker Arm Actuator System Stuck On
P2609	Intake Air Heater System Performance	P2653	"B" Rocker Arm Actuator Control Circuit Low
P2610	ECM/PCM Internal Engine Off Timer Performance	P2654	"B" Rocker Arm Actuator Control Circuit High
P2611	A/C Refrigerant Distribution Valve Control Circuit/Open	P2655	"A" Rocker Arm Actuator Control Circuit/Open
P2612	A/C Refrigerant Distribution Valve Control Circuit Low	P2656	"A" Rocker Arm Actuator System Performance or Stuck Off
P2613	A/C Refrigerant Distribution Valve Control Circuit High	P2657	"A" Rocker Arm Actuator System Stuck On
		P2658	"A" Rocker Arm Actuator Control Circuit Low
		P2659	"A" Rocker Arm Actuator Control Circuit High
		P2660	"B" Rocker Arm Actuator Control Circuit/Open
		P2661	"B" Rocker Arm Actuator System Performance or Stuck Off
		P2662	"B" Rocker Arm Actuator System Stuck On
		P2663	"B" Rocker Arm Actuator Control Circuit Low
		P2664	"B" Rocker Arm Actuator Control Circuit High
		P2665	Fuel Shutoff Valve "B" Control Circuit/Open
		P2666	Fuel Shutoff Valve "B" Control Circuit Low
		P2667	Fuel Shutoff Valve "B" Control Circuit High
		P2668	Fuel Mode Indicator Lamp Control Circuit

GENERIC OBD-II DTC CODES

Code	Description
P2669	Actuator Supply Voltage "B" Circuit /Open
P2670	Actuator Supply Voltage "B" Circuit Low
P2671	Actuator Supply Voltage "B" Circuit High
P2700	Transmission Friction Element "A" Apply Time Range/Performance
P2701	Transmission Friction Element "B" Apply Time Range/Performance
P2702	Transmission Friction Element "C" Apply Time Range/Performance
P2703	Transmission Friction Element "D" Apply Time Range/Performance
P2704	Transmission Friction Element "E" Apply Time Range/Performance
P2705	Transmission Friction Element "F" Apply Time Range/Performance
P2706	Shift Solenoid "F"
P2707	Shift Solenoid "F" Performance or Stuck Off
P2708	Shift Solenoid "F" Stuck On
P2709	Shift Solenoid "F" Electrical
P2710	Shift Solenoid "F" Intermittent
P2711	Unexpected Mechanical Gear Disengagement
P2712	Hydraulic Power Unit Leakage
P2713	Pressure Control Solenoid "D"
P2714	Pressure Control Solenoid "D" Performance or Stuck Off
P2715	Pressure Control Solenoid "D" Stuck On
P2716	Pressure Control Solenoid "D" Electrical
P2717	Pressure Control Solenoid "D" Intermittent
P2718	Pressure Control Solenoid "D" Control Circuit/Open
P2719	Pressure Control Solenoid "D" Control Circuit Range/Performance
P2720	Pressure Control Solenoid "D" Control Circuit Low
P2721	Pressure Control Solenoid "D" Control Circuit High
P2722	Pressure Control Solenoid "E"
P2723	Pressure Control Solenoid "E" Performance or Stuck Off
P2724	Pressure Control Solenoid "E" Stuck On
P2725	Pressure Control Solenoid "E" Electrical
P2726	Pressure Control Solenoid "E" Intermittent
P2727	Pressure Control Solenoid "E" Control Circuit/Open
P2728	Pressure Control Solenoid "E" Control Circuit Range/Performance
P2729	Pressure Control Solenoid "E" Control Circuit Low
P2730	Pressure Control Solenoid "E" Control Circuit High
P2731	Pressure Control Solenoid "F"
P2732	Pressure Control Solenoid "F" Performance or Stuck Off
P2733	Pressure Control Solenoid "F" Stuck On
P2734	Pressure Control Solenoid "F" Electrical
P2735	Pressure Control Solenoid "F" Intermittent
P2736	Pressure Control Solenoid "F" Control Circuit/Open
P2737	Pressure Control Solenoid "F" Control Circuit Range/Performance
P2738	Pressure Control Solenoid "F" Control Circuit Low
P2739	Pressure Control Solenoid "F" Control Circuit High
P2740	Transmission Fluid Temperature Sensor "B" Circuit
P2741	Transmission Fluid Temperature Sensor "B" Circuit Range Performance
P2742	Transmission Fluid Temperature Sensor "B" Circuit Low
P2743	Transmission Fluid Temperature Sensor "B" Circuit High
P2744	Transmission Fluid Temperature Sensor "B" Circuit Intermittent
P2745	Intermediate Shaft Speed Sensor "B" Circuit
P2746	Intermediate Shaft Speed Sensor "B" Circuit Range/Performance
P2747	Intermediate Shaft Speed Sensor "B" Circuit No Signal
P2748	Intermediate Shaft Speed Sensor "B" Circuit Intermittent
P2749	Intermediate Shaft Speed Sensor "C" Circuit
P2750	Intermediate Shaft Speed Sensor "C" Circuit Range/Performance
P2751	Intermediate Shaft Speed Sensor "C" Circuit No Signal
P2752	Intermediate Shaft Speed Sensor "C" Circuit Intermittent
P2753	Transmission Fluid Cooler Control Circuit/Open
P2754	Transmission Fluid Cooler Control Circuit Low
P2755	Transmission Fluid Cooler Control Circuit High
P2756	Torque Converter Clutch Pressure Control Solenoid
P2757	Torque Converter Clutch Pressure Control Solenoid Control Circuit Performance or Stuck Off
P2758	Torque Converter Clutch Pressure Control Solenoid Control Circuit Stuck On
P2759	Torque Converter Clutch Pressure Control Solenoid Control Circuit Electrical
P2760	Torque Converter Clutch Pressure Control Solenoid Control Circuit Intermittent
P2761	Torque Converter Clutch Pressure Control Solenoid Control Circuit/Open
P2762	Torque Converter Clutch Pressure Control Solenoid Control Circuit Range/Performance
P2763	Torque Converter Clutch Pressure Control Solenoid Control Circuit High
P2764	Torque Converter Clutch Pressure Control Solenoid Control Circuit Low
P2765	Input/Turbine Speed Sensor "B" Circuit
P2766	Input/Turbine Speed Sensor "B" Circuit Range/Performance
P2767	Input/Turbine Speed Sensor "B" Circuit No Signal
P2768	Input/Turbine Speed Sensor "B" Circuit Intermittent
P2769	Torque Converter Clutch Circuit Low
P2770	Torque Converter Clutch Circuit High
P2771	Four Wheel Drive (4WD) Low Switch Circuit
P2772	Four Wheel Drive (4WD) Low Switch Circuit Range/Performance
P2773	Four Wheel Drive (4WD) Low Switch Circuit Low
P2774	Four Wheel Drive (4WD) Low Switch Circuit High
P2775	Upshift Switch Circuit Range/Performance
P2776	Upshift Switch Circuit Low
P2777	Upshift Switch Circuit High
P2778	Upshift Switch Circuit Intermittent/Erratic
P2779	Downshift Switch Circuit Range/Performance
P2780	Downshift Switch Circuit Low
P2781	Downshift Switch Circuit High
P2782	Downshift Switch Circuit Intermittent/Erratic
P2783	Torque Converter Temperature Too High
P2784	Input/Turbine Speed Sensor "A"/"B" Correlation
P2785	Clutch Actuator Temperature Too High
P2786	Gear Shift Actuator Temperature Too High
P2787	Clutch Temperature Too High
P2788	Auto Shift Manual Adaptive Learning at Limit
P2789	Clutch Adaptive Learning at Limit

APPENDIX B

P2790 Gate Select Direction Circuit
P2791 Gate Select Direction Circuit Low
P2792 Gate Select Direction Circuit High
P2793 Gear Shift Direction Circuit
P2794 Gear Shift Direction Circuit Low
P2795 Gear Shift Direction Circuit High
P2A00 O2 Sensor Circuit Range/Performance
P2A01 O2 Sensor Circuit Range/Performance
P2A02 O2 Sensor Circuit Range/Performance
P2A03 O2 Sensor Circuit Range/Performance
P2A04 O2 Sensor Circuit Range/Performance
P2A05 O2 Sensor Circuit Range/Performance
P3400 Cylinder Deactivation System

P3000 to P4000: Displacement On-Demand Controls

P3401 Cylinder 1 Deactivation/Intake Valve Control Circuit/Open
P3402 Cylinder 1 Deactivation/Intake Valve Control Performance
P3403 Cylinder 1 Deactivation/Intake Valve Control Circuit Low
P3404 Cylinder 1 Deactivation/Intake Valve Control Circuit High
P3405 Cylinder 1 Exhaust Valve Control Circuit/Open
P3406 Cylinder 1 Exhaust Valve Control Performance
P3407 Cylinder 1 Exhaust Valve Control Circuit Low
P3408 Cylinder 1 Exhaust Valve Control Circuit High
P3409 Cylinder 2 Deactivation/Intake Valve Control Circuit/Open
P3410 Cylinder 2 Deactivation/Intake Valve Control Performance
P3411 Cylinder 2 Deactivation/Intake Valve Control Circuit Low
P3412 Cylinder 2 Deactivation/Intake Valve Control Circuit High
P3413 Cylinder 2 Exhaust Valve Control Circuit/Open
P3414 Cylinder 2 Exhaust Valve Control Performance
P3415 Cylinder 2 Exhaust Valve Control Circuit Low
P3416 Cylinder 2 Exhaust Valve Control Circuit High
P3417 Cylinder 3 Deactivation/Intake Valve Control Circuit/Open
P3418 Cylinder 3 Deactivation/Intake Valve Control Performance
P3419 Cylinder 3 Deactivation/Intake Valve Control Circuit Low
P3420 Cylinder 3 Deactivation/Intake Valve Control Circuit High
P3421 Cylinder 3 Exhaust Valve Control Circuit/Open
P3422 Cylinder 3 Exhaust Valve Control Performance
P3423 Cylinder 3 Exhaust Valve Control Circuit Low
P3424 Cylinder 3 Exhaust Valve Control Circuit High
P3425 Cylinder 4 Deactivation/Intake Valve Control Circuit/Open
P3426 Cylinder 4 Deactivation/Intake Valve Control Performance
P3427 Cylinder 4 Deactivation/Intake Valve Control Circuit Low
P3428 Cylinder 4 Deactivation/Intake Valve Control Circuit High
P3429 Cylinder 4 Exhaust Valve Control Circuit/Open
P3430 Cylinder 4 Exhaust Valve Control Performance
P3431 Cylinder 4 Exhaust Valve Control Circuit Low
P3432 Cylinder 4 Exhaust Valve Control Circuit High
P3433 Cylinder 5 Deactivation/Intake Valve Control Circuit/Open
P3434 Cylinder 5 Deactivation/Intake Valve Control Performance
P3435 Cylinder 5 Deactivation/Intake Valve Control Circuit Low
P3436 Cylinder 5 Deactivation/Intake Valve Control Circuit High
P3437 Cylinder 5 Exhaust Valve Control Circuit/Open
P3438 Cylinder 5 Exhaust Valve Control Performance
P3439 Cylinder 5 Exhaust Valve Control Circuit Low
P3440 Cylinder 5 Exhaust Valve Control Circuit High
P3441 Cylinder 6 Deactivation/Intake Valve Control Circuit/Open
P3442 Cylinder 6 Deactivation/Intake Valve Control Performance
P3443 Cylinder 6 Deactivation/Intake Valve Control Circuit Low
P3444 Cylinder 6 Deactivation/Intake Valve Control Circuit High
P3445 Cylinder 6 Exhaust Valve Control Circuit/Open
P3446 Cylinder 6 Exhaust Valve Control Performance
P3447 Cylinder 6 Exhaust Valve Control Circuit Low
P3448 Cylinder 6 Exhaust Valve Control Circuit High
P3449 Cylinder 7 Deactivation/Intake Valve Control Circuit/Open
P3450 Cylinder 7 Deactivation/Intake Valve Control Performance
P3451 Cylinder 7 Deactivation/Intake Valve Control Circuit Low
P3452 Cylinder 7 Deactivation/Intake Valve Control Circuit High
P3453 Cylinder 7 Exhaust Valve Control Circuit/Open
P3454 Cylinder 7 Exhaust Valve Control Performance
P3455 Cylinder 7 Exhaust Valve Control Circuit Low
P3456 Cylinder 7 Exhaust Valve Control Circuit High
P3457 Cylinder 8 Deactivation/Intake Valve Control Circuit/Open
P3458 Cylinder 8 Deactivation/Intake Valve Control Performance
P3459 Cylinder 8 Deactivation/Intake Valve Control Circuit Low
P3460 Cylinder 8 Deactivation/Intake Valve Control Circuit High
P3461 Cylinder 8 Exhaust Valve Control Circuit/Open
P3462 Cylinder 8 Exhaust Valve Control Performance
P3463 Cylinder 8 Exhaust Valve Control Circuit Low
P3464 Cylinder 8 Exhaust Valve Control Circuit High
P3465 Cylinder 9 Deactivation/Intake Valve Control Circuit/Open
P3466 Cylinder 9 Deactivation/Intake Valve Control Performance
P3467 Cylinder 9 Deactivation/Intake Valve Control Circuit Low
P3468 Cylinder 9 Deactivation/Intake Valve Control Circuit High
P3469 Cylinder 9 Exhaust Valve Control Circuit/Open
P3470 Cylinder 9 Exhaust Valve Control Performance
P3471 Cylinder 9 Exhaust Valve Control Circuit Low
P3472 Cylinder 9 Exhaust Valve Control Circuit High
P3473 Cylinder 10 Deactivation/Intake Valve Control Circuit/Open
P3474 Cylinder 10 Deactivation/Intake Valve Control Performance
P3475 Cylinder 10 Deactivation/Intake Valve Control Circuit Low
P3476 Cylinder 10 Deactivation/Intake Valve Control Circuit High
P3477 Cylinder 10 Exhaust Valve Control Circuit/Open
P3478 Cylinder 10 Exhaust Valve Control Performance
P3479 Cylinder 10 Exhaust Valve Control Circuit Low
P3480 Cylinder 10 Exhaust Valve Control Circuit High
P3481 Cylinder 11 Deactivation/Intake Valve Control Circuit/Open
P3482 Cylinder 11 Deactivation/Intake Valve Control Performance
P3483 Cylinder 11 Deactivation/Intake Valve Control Circuit Low
P3484 Cylinder 11 Deactivation/Intake Valve Control Circuit High
P3485 Cylinder 11 Exhaust Valve Control Circuit/Open
P3486 Cylinder 11 Exhaust Valve Control Performance
P3487 Cylinder 11 Exhaust Valve Control Circuit Low
P3488 Cylinder 11 Exhaust Valve Control Circuit High
P3489 Cylinder 12 Deactivation/Intake Valve Control Circuit/Open
P3490 Cylinder 12 Deactivation/Intake Valve Control Performance
P3491 Cylinder 12 Deactivation/Intake Valve Control Circuit Low
P3492 Cylinder 12 Deactivation/Intake Valve Control Circuit High
P3493 Cylinder 12 Exhaust Valve Control Circuit/Open
P3494 Cylinder 12 Exhaust Valve Control Performance
P3495 Cylinder 12 Exhaust Valve Control Circuit Low
P3496 Cylinder 12 Exhaust Valve Control Circuit High
P3497 Cylinder Deactivation System

APPENDIX C

Manufacturer-Specific OBD-II DTC Codes

General Motors
P1260	Fuel Pump Speed Relay Control Circuit
P1336	Crankshaft Position System Variation Not Learned
P1351	Ignition Coil Control Circuit High Voltage
P1352	Ignition Bypass Circuit High Voltage
P1361	Ignition Control (IC) Circuit Low Voltage
P1362	Ignition Bypass Circuit Low Voltage
P1374	CKP High to Low Resolution Frequency Correlation
P1380	Misfire Detected - Rough Road Data Not Available
P1381	Misfire Detected - No Communication W/Brake Control Mod
P1404	EGR Closed Position Performance
P1415	Secondary Air Injection (AIR) System Bank 1
P1416	Secondary Air Injection (AIR) System Bank 2
P1441	Evaporative Emission System Flow During Non-Purge
P1546	Air Conditioning (A/C) Clutch Relay Control Circuit
P1585	Cruise Control Inhibit Output Circuit
P1624	Customer Snapshot Requested - Data Available
P1635	5 Volt Reference Circuit
P1639	5 Volt Reference 2 Circuit
P1670	ODM has Detected a Voltage greater than 33 volts
P1689	Traction Control Delivered Torque Output Circuit
P1810	TFP Valve Position Switch Circuit
P1811	Maximum Adapt and Long Shift
P1819	Internal Mode Switch - No Start/Wrong Range
P1820	Internal Mode Switch Circuit A Low
P1822	Internal Mode Switch Circuit B High
P1823	Internal Mode Switch Circuit P Low
P1825	Internal Mode Switch - Invalid Range
P1826	Internal Mode Switch Circuit C High
P1860	TCC PWM Solenoid Circuit Electrical
P1887	TCC Release Switch Circuit

Ford Motor Company
P1000	OBD-II Monitor Testing Incomplete
P1001	KOER Test Cannot Be Completed
P1039	Vehicle Speed Signal Missing or Improper
P1051	Brake Switch Signal Missing or Improper
P1100	Mass Air Flow Sensor Intermittent
P1101	Mass Air Flow Sensor out of Self-Test Range
P1112	Intake Air Temperature Sensor Intermittent
P1114	Intake Air Temperature 2 Circuit Low Input
P1115	Intake Air Temperature 2 Circuit High Input
P1116	Engine Coolant Temperature Sensor out of Self-Test Range
P1117	Engine Coolant Temperature Sensor Intermittent
P1120	Throttle Position Sensor out of Range
P1121	Throttle Position Sensor Inconsistent with Mass Air Flow Sensor
P1124	Throttle Position Sensor out of Self-Test Range
P1125	Throttle Position Sensor Intermittent
P1127	Heated Oxygen Sensor Heater not on During KOER Test - Exhaust Not Warm Enough
P1129	Upstream O2 Sensors Swapped Bank To Bank (HO2S-11-21)
P1130	Lack of Upstream Heated Oxygen Sensor Switch - Adaptive Fuel Limit Bank No. 1
P1131	Lack of Upstream Heated Oxygen Sensor Switch - Sensor Indicates Lean Bank No. 1
P1132	Lack of Upstream Heated Oxygen Sensor Switch - Sensor Indicates Rich Bank No. 1
P1135	Ignition Switch Signal Missing or Improper
P1137	Lack of Downstream Heated Oxygen Sensor Switch - Sensor Indicates Lean Bank No. 1
P1138	Lack of Downstream Heated Oxygen Sensor Switch - Sensor Indicates Rich Bank No. 1
P1150	Lack of Upstream Heated Oxygen Sensor Switch - Adaptive Fuel Limit Bank No. 2
P1151	Lack of Upstream Heated Oxygen Sensor Switch - Sensor Indicates Lean Bank No. 2
P1152	Lack of Upstream Heated Oxygen Sensor Switch - Sensor Indicates Rich Bank No. 2
P1157	Lack of Downstream Heated Oxygen Sensor Switch - Sensor Indicates Lean Bank No. 2
P1158	Lack of Downstream Heated Oxygen Sensor Switch - Sensor Indicates Rich Bank No. 2
P1168	FRP Sensor In Range But Low
P1169	FRP Sensor In Range But High
P1180	Fuel Delivery System Low
P1181	Fuel Delivery System High
P1183	EOT Sensor Circuit Malfunction
P1184	EOT Sensor out of Range
P1220	Series Throttle Control fault
P1224	Throttle Position Sensor B out of Self-Test Range
P1229	Supercharger Intercooler Pump Not Working
P1230	Open Power to Fuel Pump Circuit
P1231	High Speed Fuel Pump Relay Activated
P1232	Low Speed Fuel Pump Primary Circuit Failure
P1233	Fuel Pump Driver Module Off-line
P1234	Fuel Pump Driver Module Off-line
P1235	Fuel Pump Control out of Range
P1236	Fuel Pump Control out of Range

APPENDIX C

Code	Description
P1237	Fuel Pump Secondary Circuit Fault
P1238	Fuel Pump Secondary Circuit Fault
P1244	Generator Load Input Low
P1245	Generator Load Input High
P1246	Generator Load Input Failed
P1250	Lack of Power to FPRC Solenoid
P1260	Theft Detected - Engine Disabled
P1270	Engine RPM or Vehicle Speed Limiter Reached
P1285	Cylinder Head Over Temperature Sensed
P1288	Cylinder Head Temperature Sensor out of Self-Test Range
P1289	Cylinder Head Temperature Sensor Signal Greater Than Self-Test Range
P1290	Cylinder Head Temperature Sensor Signal Less Than Self-Test Range
P1299	Cylinder Head Temperature Sensor Detected Engine Overheating Condition
P1309	Misfire Detection Monitor not Enabled
P1336	CKP and/or CMP Input Signal to PCM Concerns
P1351	Ignition Diagnostic Monitor Circuit Input Fault
P1352	Ignition Coil A - Primary Circuit Fault
P1353	Ignition Coil B - Primary Circuit Fault
P1354	Ignition Coil C - Primary Circuit Fault
P1355	Ignition Coil D - Primary Circuit Fault
P1356	Loss of Ignition Diagnostic Module Input to PCM
P1358	Ignition Diagnostic Monitor Signal out of Self-Test Range
P1359	Spark Output Circuit Fault
P1364	Ignition Coil Primary Circuit Fault
P1380	Variable Cam Timing Solenoid "A" Circuit Malfunction
P1381	Variable Cam Timing Over-Advanced (Bank 1)
P1383	Variable Cam Timing Over-Retarded (Bank 1)
P1390	Octane Adjust out of Self-Test Range
P1400	Differential Pressure Feedback Electronic Sensor Circuit Low Voltage
P1401	Differential Pressure Feedback Electronic Sensor Circuit High Voltage
P1403	Differential Pressure Feedback Electronic Sensor Hoses Reversed
P1405	Differential Pressure Feedback Electronic Sensor Circuit Upstream Hose
P1406	Differential Pressure Feedback Electronic Sensor Circuit Downstream Hose
P1407	EGR No Flow Detected
P1408	EGR Flow out of Self-Test Range
P1409	EGR Vacuum Regulator Circuit Malfunction
P1410	EGR Barometric Pressure Sensor VREF Voltage
P1411	Secondary Air Is Not Being Diverted
P1413	Secondary Air Injection System Monitor Circuit Low Voltage
P1414	Secondary Air Injection System Monitor Circuit High Voltage
P1432	THTRC Circuit Failure
P1436	A/C Evaporator Temperature Circuit Low Input
P1437	A/C Evaporator Temperature Circuit High Input
P1442	Secondary Air Injection System Monitor Circuit High Voltage
P1443	Evaporative Emission Control System - Vacuum System - Purge Control Solenoid or Purge Control Valve fault
P1444	Purge Flow Sensor Circuit Input Low
P1445	Purge Flow Sensor Circuit Input High
P1450	Inability of Evaporative Emission Control System to Bleed Fuel Tank
P1451	EVAP Control System Canister Vent Solenoid Circuit Malfunction
P1452	Inability of Evaporative Emission Control System to Bleed Fuel Tank
P1455	Substantial Leak or Blockage in Evaporative Emission Control System
P1460	Wide Open Throttle Air Conditioning Cutoff Circuit Malfunction
P1461	Air Conditioning Pressure Sensor Circuit Low Input
P1462	Air Conditioning Pressure Sensor Circuit High Input
P1463	Air Conditioning Pressure Sensor Insufficient Pressure Change
P1464	ACCS to PCM High During Self-Test
P1469	Low Air Conditioning Cycling Period
P1473	Fan Secondary High with Fans Off
P1474	Low Fan Control Primary Circuit
P1477	MFC Primary Circuit Failure
P1479	High Fan Control Primary Circuit
P1480	Fan Secondary Low with Low Fans On
P1481	Fan Secondary Low with High Fans On
P1483	Power to Cooling Fan Exceeded Normal Draw
P1484	Variable Load Control Module Pin 1 Open
P1500	Vehicle Speed Sensor (VSS) Intermittent
P1501	Programmable Speedometer & Odometer Module/Vehicle Speed Sensor (VSS) Intermittent Circuit Failure
P1502	Invalid or Missing Vehicle Speed Message or Brake Data
P1504	Intake Air Control Circuit malfunction
P1505	Idle Air Control System at Adaptive Clip
P1506	Idle Air Control Over-Speed Error
P1507	Idle Air Control Under-Speed Error
P1512	Intake Manifold Runner Control Stuck Closed
P1513	Intake Manifold Runner Control Stuck Closed
P1516	Intake Manifold Runner Control Input Error
P1517	Intake Manifold Runner Control Input Error
P1518	Intake Manifold Runner Control Fault - Stuck Open
P1519	Intake Manifold Runner Control Fault - Stuck Closed
P1520	Intake Manifold Runner Control Circuit fault
P1530	Open or Short to A/C Compressor Clutch Circuit
P1537	Intake Manifold Runner Control Stuck Open
P1538	Intake Manifold Runner Control Stuck Open
P1539	Power to A/C Compressor Clutch Circuit Exceeded Normal Current Draw
P1549	Intake Manifold Temperature Valve Vacuum Actuator Connection
P1550	Power Steering Pressure Sensor out of Self-Test Range
P1572	Brake Pedal Switch Circuit
P1605	PCM Keep Alive Memory Test Error
P1625	Voltage to Vehicle Load Control Module Fan Circuit Not Detected
P1626	Voltage to Vehicle Load Control Module Circuit Not Detected
P1633	Keep Alive Power Voltage Too Low
P1635	Tire/Axle out of Acceptable Range
P1636	Inductive Signature Chip Communication Error
P1639	Vehicle ID Block not Programmed or Corrupt
P1640	DTCs Available In Another Module
P1650	Power Steering Pressure Switch out of Self-Test Range
P1651	Power Steering Pressure Switch Signal Malfunction
P1700	Transmission System Problems
P1701	Reverse Engagement Error
P1702	Transmission System Problems
P1703	Brake On/Off Switch out of Self-Test Range
P1704	Transmission System Problems
P1705	Transmission Range Sensor out of Self-Test Range
P1709	Park or Neutral Position Switch out of Self-Test Range
P1710	Transmission System Problems
P1711	Transmission Fluid Temperature Sensor out of Self-Test Range
P1713	Transmission system problems
P1714	Transmission System Problems
P1715	Transmission System Problems
P1716	Transmission System Problems
P1717	Transmission System Problems
P1718	Transmission System Problems
P1719	Transmission System Problems
P1720	Transmission System Problems
P1729	4x4 Low Switch Error
P1740	Transmission System Problems
P1741	Torque Converter Clutch Control Error
P1742	Torque Converter Clutch Solenoid Faulty
P1743	Torque Converter Clutch Solenoid Faulty
P1744	Torque Converter Clutch System Stuck in Off Position
P1745	Transmission System Problems
P1746	Electronic Pressure Control Solenoid - Open Circuit
P1747	Electronic Pressure Control Solenoid - Short Circuit
P1749	Electronic Pressure Control Solenoid Failed Low
P1751	Shift Solenoid No. 1 Performance
P1754	Coast Clutch Solenoid Circuit malfunction
P1756	Shift Solenoid No. 2 Performance
P1760	Transmission System Problems
P1761	Shift Solenoid No. 3 Performance
P1762	Transmission System Problems
P1767	Transmission System Problems
P1780	Transmission Control Switch Circuit is out of Self-Test Range
P1781	4x4 Low Switch is out of Self-Test Range

OBD-II DTC CODES

P1783	Transmission Over-Temperature Condition	P1681	No I/P Cluster CCD/J1850 Messages Received
P1784	Transmission System Problems	P1682	Charging System Voltage Too Low
P1785	Transmission System Problems	P1683	Speed Control Power Relay Or Speed Control 12 Volt Driver Circuit
P1786	Transmission System Problems	P1684	Battery Disconnected within Last 50 Starts
P1787	Transmission System Problems	P1685	Skim Invalid Key
P1788	Transmission System Problems	P1686	No SKIM Bus Message Received
P1789	Transmission System Problems	P1687	No Cluster Bus Message
P1900	Transmission System Problems	P1688	Internal Fuel Injection Pump Controller Failure
		P1689	No Communication Between ECM & Injection Pump Module

Chrysler/Dodge/Jeep

P1192	Inlet Air Temperature Circuit Low	P1690	Fuel Injection Pump CKP Sensor Does Not Agree With ECM CKP Sensor
P1193	Inlet Air Temperature Circuit High	P1691	Fuel Injection Pump Controller Calibration Failure
P1195	1/1 O2 Sensor Slow During Catalyst Monitor	P1693	DTC Detectedin ECM or PCM
P1196	2/1 O2 Sensor Slow During Catalyst Monitor	P1694	No CCD Messages Received from ECM
P1197	1/2 O2 Sensor Slow During Catalyst Monitor	P1695	No CCD/J185O Message from BCM
P1198	Radiator Temperature Sensor Volts Too High	P1696	PCM Failure EEPROM Write Denied
P1199	Radiator Temperature Sensor Volts Too Low	P1697	PCM Failure SRI Mile Not Stored
P1281	Engine is Cold Too Long	P1698	No CCD Messages Received from PCM
P1282	Fuel Pump Relay Control Circuit	P1719	Skip Shift Solenoid Circuit
P1283	Idle Select Signal Invalid	P1740	TCC or OD Solenoid Performance
P1284	Fuel Injection Pump Battery Voltage out of Range	P1756	Governor Pressure Not Equal to Target at 15-20 psi
P1285	Fuel Injection Pump Controller Always On	P1757	Governor Pressure Above 3 psi When Request Is 0 psi
P1286	Accelerator Pedal Position Sensor Supply Voltage Too High	P1762	Governor Pressure Sensor Offset Improper Voltage
P1287	Fuel Injection Pump Controller Supply Voltage Low	P1763	Governor Pressure Sensor Voltage Too High
P1288	Intake Manifold Short Runner Solenoid Circuit	P1764	Governor Pressure Sensor Voltage Too Low
P1289	Manifold Tune Valve Solenoid Circuit	P1765	Trans 12 Volt Supply Relay Control Circuit
P1290	CNG Fuel Pressure Too High	P1899	Park/Neutral Position Switch Stuck in Park or in Gear
P1291	No Temp Rise Seen From Fuel Heaters		
P1292	CNG Pressure Sensor Voltage Too High		
P1293	CNG Pressure Sensor Voltage Too Low		

Toyota

P1294	Target Idle Not Reached	P1100	BARO Sensor Circuit
P1295	No 5 Volts to TP Sensor	P1120	Accelerator Pedal Position Sensor Circuit
P1296	No 5 Volts to MAP Sensor	P1121	Accelerator Pedal Position Sensor Range/Performance Problem
P1297	No Change in MAP From Start To Run	P1125	Throttle Control Motor Circuit
P1298	Lean Operation at Wide-Open Throttle	P1126	Magnetic Clutch Circuit
P1299	Vacuum Leak Found (IAC Fully Seated)	P1127	ETCS Actuator Power Source Circuit
P1388	Auto Shutdown (ASD) Relay Control Circuit	P1128	Throttle Control Motor Lock
P1389	No Auto Shutdown (ASD) Relay Output Voltage At PCM	P1129	Electric Throttle Control System
P1390	Timing Belt Skipped One Tooth or More	P1130	Air/Fuel Sensor Circuit Range/Performance (Bank 1 Sensor 1)
P1391	Intermittent Loss of CMP or CKP	P1133	Air/Fuel Sensor Circuit Response (Bank 1 Sensor 1)
P1398	Misfire Adapter Numerator at Limit	P1135	Air/Fuel Sensor Heater Circuit Response (Bank 1 Sensor 1)
P1399	Wait to Start Lamp Circuit	P1150	Air/Fuel Sensor Circuit Range/Performance (Bank 1 Sensor 2)
P1403	No 5 Volts to EGR Sensor	P1153	Air/Fuel Sensor Circuit Response (Bank 1 Sensor 2)
P1475	Aux. 5 Volt Output Too High	P1155	Air/Fuel Sensor Heater Circuit (Bank 1 Sensor 2)
P1476	Too Little Secondary Air	P1200	Fuel Pump Relay Circuit
P1477	Too Much Secondary Air	P1300	Igniter Circuit Malfunction - No. 1
P1478	Battery Temp Sensor Volts out of Limit	P1310	Igniter Circuit Malfunction - No. 2
P1479	Transmission Fan Relay Circuit	P1335	No Crankshaft Position Sensor Signal - Engine Running
P1480	PCV Solenoid Valve	P1349	VVT System
P1482	Catalyst Temperature Sensor Circuit Shorted Low	P1400	Sub-Throttle Position Sensor
P1483	Catalyst Temperature Sensor Circuit Shorted High	P1401	Sub-Throttle Position Sensor Range/Performance Problem
P1484	Catalytic Converter Overheat Detected	P1405	Turbo Pressure Sensor Circuit
P1485	Air Injection Solenoid Circuit	P1406	Turbo Pressure Sensor Range/Performance Problem
P1486	Evap. Leak Monitor Pinched Hose	P1410	EGR Valve Position Sensor Circuit Malfunction
P1487	High Speed Radiator Fan CTRL Relay Circuit	P1411	EGR Valve Position Sensor Circuit Range/Performance
P1488	Auxiliary 5 Volt Supply Output Too Low	P1500	Starter Signal Circuit
P1489	High Speed Fan CTRL Relay Circuit	P1510	Boost Pressure Control Circuit
P1490	Low Speed Fan CTRL Relay Circuit	P1511	Boost Pressure Low
P1491	Radiator Fan Control Relay Circuit	P1512	Boost Pressure High
P1492	Battery Temperature Sensor Voltage Too High	P1520	Stop Lamp Switch Signal Malfunction
P1493	Battery Temperature Sensor Voltage Too Low	P1565	Cruise Control Main Switch Circuit
P1494	Leak Detection Pump Switch or Mechanical Fault	P1600	ECM BATT Malfunction
P1495	Leak Detection Pump Solenoid Circuit	P1605	Knock Control CPU
P1496	5 Volt Supply Output Too Low	P1630	Traction Control System
P1498	High speed Radiator Fan Ground CTRL Rely Circuit	P1633	ECM
P1594	Charging System Voltage Too High	P1652	Idle Air Control Valve Control Circuit
P1595	Speed Control Solenoid Circuits	P1656	OCV Circuit
P1596	Speed Control Switch Always High	P1658	Wastegate Valve Control Circuit
P1597	Speed Control Switch Always Low	P1661	EGR Circuit
P1598	A/C Pressure Sensor Volts Too High	P1662	EGR Bypass Valve Control Circuit
P1599	A/C Pressure Sensor Volts Too Low	P1780	Park/Neutral Position Switch Malfunction (Only For A/T)
P1602	PCM Not Programmed	P1875	4WD Low Switch Circuit Malfunction
P1680	Clutch Released Switch Circuit		

AUTOMOTIVE DIAGNOSTIC SYSTEMS

APPENDIX C

Honda
P1106	Barometric Pressure Circuit Range/Performance
P1107	Barometric Pressure Circuit Low Input
P1108	Barometric Pressure Circuit High Input
P1121	Throttle Position Lower than Expected
P1122	Throttle Position Higher than Expected
P1128	MAP Lower than Expected
P1129	MAP Higher than Expected
P1149	Primary HO2S (Sensor 1) Circuit Range/Performance Problem
P1162	Primary HO2S (No. 1) Circuit Malfunction
P1163	Primary HO2S (No. 1) Circuit Slow Response
P1164	Primary HO2S (No. 1) Circuit Range/Performance
P1165	Primary HO2S (No. 1) Circuit Range/Performance
P1166	Primary HO2S (No. 1) Heater System Electrical
P1167	Primary HO2S (No. 1) Heater System
P1168	Primary HO2S (No. 1) Label Low Input
P1169	Primary HO2S (No. 1) Label High Input
P1253	VTEC System Malfunction
P1257	VTEC System Malfunction
P1258	VTEC System Malfunction
P1259	VTEC System Malfunction
P1297	Electrical Load Detector Circuit Low Input
P1298	Electrical Load Detector Circuit High Input
P1300	Multiple Cylinder Misfire Detected
P1336	CSF Sensor Intermittent Interruption
P1337	CSF Sensor No Signal
P1359	CKP/TDC Sensor Connector Disconnection
P1361	Intermittent Interruption In TDC 1 Sensor Circuit
P1362	No Signal In TDC 1 Sensor Circuit
P1366	Intermittent Interruption In TDC 2 Sensor Circuit
P1367	No Signal In TDC 2 Sensor Circuit
P1381	Cylinder Position Sensor Intermittent Interruption
P1382	Cylinder Position Sensor No Signal
P1456	EVAP Emission Control System Leak Detected (Fuel Tank System)
P1457	EVAP Emission Control System Leak Detected (Control Canister System)
P1459	EVAP Emission Purge Flow Switch Malfunction
P1486	Thermostat Range/Performance Problem
P1491	EGR Valve Lift Insufficient Detected
P1498	EGR Valve Lift Sensor High Voltage
P1508	IAC Valve Circuit Failure
P1509	IAC Valve Circuit Failure
P1519	Idle Air Control Valve Circuit Failure
P1607	ECM/PCM Internal Circuit Failure A
P1655	SEAF/SEFA/TMA/TMB Signal Line Failure
P1656	Automatic Transaxle
P1660	Automatic Transaxle FI Signal A Circuit Failure
P1676	FPTDR Signal Line Failure
P1678	FPTDR Signal Line Failure
P1681	Automatic Transaxle FI Signal A Low Input
P1682	Automatic Transaxle FI Signal A High Input
P1686	Automatic Transaxle FI Signal B Low Input
P1687	Automatic Transaxle FI Signal B High Input
P1705- P1891	Automatic Transaxle Concerns

Hyundai
P1100	Manifold Absolute Pressure (MAP) Sensor Malfunction (Open/Short)
P1102	Manifold Absolute Pressure (MAP) Sensor Malfunction - Low Voltage
P1103	Manifold Absolute Pressure (MAP) Sensor Malfunction - High Voltage
P1147	ETS Sub Accel Position Sensor 1 Malfunction
P1151	ETS Main Accel Position Sensor 2 Malfunction
P1155	ETS Limp Home Valve
P1159	Variable Intake Motor Malfunction
P1171	Electronic Throttle System Open
P1172	Electronic Throttle System Motor Current
P1173	Electronic Throttle System Rationality Malfunction
P1174	Electronic Throttle System #1 Close Malfunction
P1175	Electronic Throttle System #2 Close Malfunction
P1176	ETS Motor Open/Short #1
P1176	ETS Motor Open/Short #2
P1178	ETS Motor Battery Voltage Open
P1330	Spark Timing Adjust Malfunction
P1521	Power Steering Switch Malfunction
P1607	Electronic Throttle System Communication Error
P1614	Electronic Throttle System Module Malfunction
P1632	Traction Control System Malfunction

Kia
P1115	Engine Coolant Temperature Signal from ECM to TCM
P1121	Throttle Position Sensor Signal Malfunction from ECM to TCM
P1170	Front Heated Oxygen Sensor Stuck
P1195	EGR Pressure Sensor (1.6L) or Boost Sensor (1.8L) Open or Short
P1196	Ignition Switch "Start" Open or Short (1.6L)
P1250	Pressure Regulator Control Solenoid Valve Open or Short
P1252	Pressure Regulator Control Solenoid Valve No. 2 Circuit Malfunction
P1307	Chassis Acceleration Sensor Signal Malfunction
P1308	Chassis Acceleration Sensor Signal Low
P1309	Chassis Acceleration Sensor Signal High
P1345	No SGC Signal (1.6L)
P1386	Knock Sensor Control Zero Test
P1402	EGR Valve Position Sensor Open or Short
P1449	Canister Drain Cut Valve Open or Short (1.8L)
P1450	Excessive Vacuum Leak
P1455	Fuel Tank Sending Unit Open or Short (1.8L)
P1457	Purge Solenoid Valve Low System Malfunction
P1458	A/C Compressor Control Signal Malfunction
P1485	EGR Solenoid Valve Vacuum Open or Short
P1486	EGR Solenoid Valve Vent Open or Short
P1487	EGR Boost Sensor Solenoid Valve Open or Short
P1496	EGR Stepper Motor Malfunction - Circuit 1 (1.8L)
P1497	EGR Stepper Motor Malfunction - Circuit 2 (1.8L)
P1498	EGR Stepper Motor Malfunction - Circuit 3 (1.8L)
P1499	EGR Stepper Motor Malfunction - Circuit 4 (1.8L)
P1500	No Vehicle Speed Signal to TCM
P1505	Idle Air Control Valve Opening Coil Voltage Low
P1506	Idle Air Control Valve Opening Coil Voltage High
P1507	Idle Air Control Valve Closing Coil Voltage Low
P1508	Idle Air Control Valve Closing Coil Voltage High
P1523	VICS Solenoid Valve
P1586	A/T-M/T Codification
P1608	PCM Malfunction
P1611	MIL Request Circuit Voltage Low
P1614	MIL Request Circuit Voltage High
P1624	MIL Request Signal from TCM to ECM
P1631	Alternator "T" Open or No Power Output (1.8L)
P1632	Battery Voltage Detection Circuit for Alternator Regulator (1.8L)
P1633	Battery Overcharge
P1634	Alternator "B" Open (1.8L)
P1693	MIL Circuit Malfunction
P1743	Torque Converter Clutch Solenoid Valve Open or Short
P1794	Battery or Circuit Failure
P1795	4WD Switch Signal Malfunction
P1797	P or N Range Signal or Clutch Pedal Position Switch Open or Short

Mazda
P1100	MAF Intermittant
P1101	MAF out of Range
P1116	ECT Sensor out of Range
P1117	ECT Intermittent
P1120	TPS out of Range Low
P1124	TPS out of Self-Test Range
P1125	Tandem Throttle Position Sensor
P1129	Downstream O2 Sensors Swapped Bank to Bank (HO2S-12-22)
P1130	Heated O2 Sensor (HO2S) 11 at Adaptive Limit
P1131	HO2S 11 Indicates Lean
P1132	HO2S 11 Indicates Rich
P1260	Theft Detected - Engine Disabled
P1270	Vehicle Speed Limiter Reached
P1390	Octane Adjust out of Range
P1400	DPFE Sensor Low Voltage
P1401	DPFE Sensor High Voltage
P1403	DPFE Hoses Reversed

OBD-II DTC CODES

Code	Description
P1405	DPFE Upstream Hose Off or Plugged
P1406	DPFE Downstream Hose Off or Plugged
P1407	EGR No Flow Detected
P1408	EGR out of Self Test Range
P1409	EGR Vacuum Regulator Solenoid Circuit Malfunction
P1443	Evaporative Emission Control System
P1444	Purge Flow Sensor Low Input
P1445	Purge Flow Sensor High Input
P1460	WOT A/C Cutoff Circuit Malfunction
P1500	Vehicle Speed Sensor (VSS) Intermittant
P1505	Idle Air Control at Adaptive Clip
P1506	Idle Air Control System Overspeed Error
P1507	Idle Air Control System Underspeed Error
P1605	Powertrain Control Module
P1650	Power Steering Pressure Switch out of Range
P1651	Power Steering Pressure Switch Input Malfunction
P1701	Transmission Solenoid Malfunction
P1703	Brake On/Off Switch
P1705	Transmisson Range Sensor
P1709	Park/Neutral Position Switch out of Range
P1711	Transmission Fluid Temperature Sensor
P1729	4x4 Low Switch Malfunction
P1746	EPC Solenoid Failed Low
P1747	EPC Solenoid Short Circuit
P1749	EPC Solenoid Open Circuit
P1751	Shift Solenoid #1 (SS1)
P1754	Coast Clutch Solenoid
P1756	Shift Solenoid #2 (SS2)
P1761	Shift Solenoid #3 (SS3)
P1780	Transmission Control Switch
P1781	4x4 Switch out of Range
P1783	Transmission Over-Temperature Condition

Nissan

Code	Description
P1105	MAP/BARO Pressure Switch Solenoid Valve
P1126	Thermostat Function
P1130	Swirl Control Valve Control Solenoid Valve
P1148	Closed Loop Control (Bank 1)
P1165	Swirl Control Valve Control Vacuum Switch
P1168	Closed Loop Control (Bank 2)
P1320	Ignition Signal
P1211	ABS/TCS Control Unit
P1212	ABS/TCS Communication Line
P1217	Engine Over-Temperature (overheat)
P1320	Ignition Signal
P1335	Crankshaft Position Sensor (REF)
P1336	Crankshaft Position Sensor (CKPS)
P1400	EGRC Solenoid Valve
P1401	EGR Temperature Sensor
P1402	EGR System
P1440	EVAP Control System Small Leak
P1441	Vacuum Cut Valve Bypass Valve
P1444	Canister Purge Volume Control Solenoid Valve
P1445	EVAP Canister Purge Volume Control Valve
P1446	EVAP Canister Vent Control Valve (Closed)
P1447	EVAP Control System Purge Flow Monitoring
P1448	EVAP Canister Vent Control Valve (Open)
P1464	Fuel Level Sensor Circuit (Ground Signal)
P1490	Vacuum Cut Valve Bypass Valve (Circuit)
P1491	Vacuum Cut Valve Bypass Valve
P1492	EVAP Canister Purge Control/Solenoid Valve (Circuit)
P1493	EVAP Canister Purge Control Valve/Solenoid Valve
P1550	TCC Solenoid Valve
P1605	A/T Diagnostic Communication Line
P1705	Throttle Position Sensor Circuit A/T
P1706	Park/Neutral Position (PNP) Switch
P1760	Overrun Clutch Solenoid Valve (Circuit)

Volkswagen

Code	Description
P1102	O2S Heating Circuit Bank 1 Sensor 1 Voltage Too Low/Air Leak
P1105	O2S Heating Circuit Bank 1 Sensor 2 Short to Positive
P1107	O2S Heating Circuit Bank 2 Sensor 1 Short to Positive
P1110	O2S Heating Circuit Bank 2 Sensor 2 Short to Positive
P1113	O2S Sensor Heater Resistance Too High Bank 1 Sensor 1
P1115	O2S Sensor Heater Circuit Short to Ground Bank 1 Sensor 1
P1116	O2S Sensor Heater Circuit Open Bank 1 Sensor 1
P1117	O2S Sensor Heater Circuit Short to Ground Bank 1 Sensor 2
P1118	O2S Sensor Heater Circuit Open Bank 1 Sensor 2
P1127	Long Term Fuel Trim B1 System Too Rich
P1128	Long Term Fuel Trim B1 System Too Lean
P1129	Long Term Fuel Trim B2 System Too Rich
P1130	Long Term Fuel Trim B2 System Too Lean
P1136	Long Term Fuel Trim Add. Fuel B1 System Too Lean
P1137	Long Term Fuel Trim Add. Fuel B1 System Too Rich
P1138	Long Term Fuel Trim Add. Fuel B2 System Too Lean
P1139	Long Term Fuel Trim Add. Fuel B2 System Too Rich
P1141	Load Calculation Cross Check Range/Performance
P1144	Mass Air Flow Sensor Open/Short to Ground
P1145	Mass Air Flow Sensor Short to Positive
P1146	Mass Air Flow Sensor Supply Voltage
P1155	Manifold Absolute Pressure Sensor Short to Positive
P1156	Manifold Absolute Pressure Sensor Open/Short to Ground
P1157	Manifold Absolute Pressure Sensor Supply Voltage
P1160	Intake Air Temperature Sensor Short to Ground
P1161	Intake Air Temperature Sensor Open/Short to Positive
P1162	Intake Air Temperature Sensor Short to Ground
P1163	Fuel Temperature Sensor Open/Short to Positive
P1164	Fuel Temperature Sensor Implausible Signal
P1171	Throttle Actuation Potentiometer Sign. 2 Range/Performance
P1172	Throttle Actuation Potentiometer Sign. 2 Signal Too Low
P1173	Throttle Actuation Potentiometer Sign. 2 Signal Too High
P1176	Rear O2S Correction
P1177	O2 Correction Behind Catalyst B1 Limit Attained
P1196	O2S Heater Circuit Bank 1 Sensor 1 Electrical Malfunction
P1197	O2S Heater Circuit Bank 2 Sensor 1 Electrical Malfunction
P1198	O2S Heater Circuit Bank 1 Sensor 2 Electrical Malfunction
P1199	O2S Heater Circuit Bank 2 Sensor 2 Electrical Malfunction
P1213	Injector Circuit Cylinder 1 Short to Positive
P1214	Injector Circuit Cylinder 2 Short to Positive
P1215	Injector Circuit Cylinder 3 Short to Positive
P1216	Injector Circuit Cylinder 4 Short to Positive
P1217	Injector Circuit Cylinder 5 Short to Positive
P1218	Injector Circuit Cylinder 6 Short to Positive
P1225	Injector Circuit Cylinder 1 Short to Ground
P1226	Injector Circuit Cylinder 2 Short to Ground
P1227	Injector Circuit Cylinder 3 Short to Ground
P1228	Injector Circuit Cylinder 4 Short to Ground
P1229	Injector Circuit Cylinder 5 Short to Ground
P1230	Injector Circuit Cylinder 6 Short to Ground
P1237	Injector Circuit Open Cylinder 1
P1238	Injector Circuit Open Cylinder 2
P1239	Injector Circuit Open Cylinder 3
P1240	Injector Circuit Open Cylinder 4
P1241	Injector Circuit Open Cylinder 5
P1242	Injector Circuit Open Cylinder 6
P1245	Needle Lift Sensor Short to Ground
P1246	Needle Lift Implausible Signal
P1247	Needle Lift Sensor Open/Short to Positive
P1248	Start of Cold Start Injector Control Difference
P1251	Start of Cold Start Injector Short to Positive
P1252	Start of Cold Start Injector Open/Short to Ground
P1255	Engine Coolant Temperature Sensor Short to Ground
P1256	Engine Coolant Temperature Sensor Open/Short to Positive
P1300	Misfire Detected, Fuel Level Too Low
P1250	Fuel Level Too Low
P1325	Cyl. 1 Knock Control Limit Attained
P1326	Cyl. 2 Knock Control Limit Attained
P1327	Cyl. 3 Knock Control Limit Attained
P1328	Cyl. 4 Knock Control Limit Attained
P1329	Cyl. 5 Knock Control Limit Attained
P1330	Cyl. 6 Knock Control Limit Attained
P1336	Engine Torque Adaption at Limit
P1337	CMP Sensor Bank 1 Short to Ground
P1338	CMP Sensor Bank 1 Open Circuit or Short to Positive
P1340	CKP/CMP Sensor Signals out of Sequence

APPENDIX C

Code	Description
P1341	Ignition Coil Output Stage 1 Short to Ground
P1343	Ignition Coil Output Stage 2 Short to Ground
P1345	Ignition Coil Output Stage 3 Short to Ground
P1354	Modulating Piston Displacement Sensor Electrical Circuit Malfunction
P1386	Internal Control Module Knock Control Error
P1387	Control Unit Internal Altitude Sensor
P1391	CMP Sensor Bank 2 Short to Ground
P1392	CMP Sensor Bank 2 Open Circuit/Short to Positive
P1393	Ignition Coil Power Output Stage 1 Malfunction
P1394	Ignition Coil Power Output Stage 2 Malfunction
P1395	Ignition Coil Power Output Stage 3 Malfunction
P1401	EGR Valve Power Stage Short To Ground
P1402	EGR Vacuum Regulator Solenoid Valve Short to Positive
P1403	EGR System Control Difference
P1407	EGR Temperature Sensor Signal Too Low
P1408	EGR Temperature Sensor Signal Too High
P1410	Tank Ventilation Valve Circuit Short to B+
P1420	Secondary Air Injection Control Module Electrical Malfunction
P1421	Secondary Air Injection Valve Circuit Short to Ground
P1422	Secondary Air Injection Valve Circuit Short to B+
P1424	Secondary Air Injection System Bank Leak Detected
P1425	Tank Ventilation Valve Short to Ground
P1426	Tank Ventilation Valve Open Circuit
P1432	Secondary Air Injection Valve Open
P1433	Secondary Air Injection Pump Relay Circuit Open
P1434	Secondary Air Injection Pump Relay Circuit Short to Positive
P1435	Secondary Air Injection Pump Relay Circuit Short to Ground
P1436	Secondary Air Injection Pump Relay Circuit Electrical Malfunction
P1440	EGR Valve Power Stage Open
P1441	EGR Vacuum Regulator Solenoid Valve Open/Short to Ground
P1450	Secondary Air Injection System Circuit Short to Positive
P1451	Secondary Air Injection Circuit Short to Ground
P1452	Secondary Air Injection System Circuit Open
P1471	EVAP Control System LDP Circuit Short to Positive
P1472	EVAP Control System LDP Circuit Short to Ground
P1473	EVAP Control System LDP Open Circuit
P1475	EVAP Control System LDP Malfunction/Signal Circuit Open
P1476	EVAP Control System LDP Malfunction/Insufficient Vacuum
P1477	EVAP Control System LDP Malfunction
P1478	EVAP Control System LDP Clamped Tube Detected
P1500	Fuel Pump Relay Electrical Circuit Malfunction
P1501	Fuel Pump Relay Circuit Short to Ground
P1502	Fuel Pump Relay Circuit Short to Positive
P1505	Closed Throttle Position Does Not Close/Open Circuit
P1506	Closed Throttle Position Switch Does Not Open/Short to Ground
P1512	Intake Manifold Changeover Valve Circuit Short to Positive
P1515	Intake Manifold Changeover Valve Circuit Short to Ground
P1516	Intake Manifold Changeover Valve Circuit Open
P1519	Intake Camshaft Control Bank 1 Malfunction
P1522	Intake Camshaft Control Bank 2 Malfunction
P1537	Fuel Cut-off Valve Incorrect Function
P1538	Fuel Cut-off Valve Open/Short to Ground
P1539	Clutch Pedal Switch Signal Fault
P1540	VSS Signal Too High
P1541	Fuel Pump Relay Circuit Open
P1542	Throttle Actuation Potentiometer Range/Performance
P1543	Throttle Actuation Potentiometer Signal Too Low
P1544	Throttle Actuation Potentiometer Signal Too High
P1545	Throttle Position Control Malfunction
P1546	Wastegate Bypass Regulator Valve Short to Positive
P1547	Wastegate Bypass Regulator Valve Short to Ground
P1548	Wastegate Bypass Regulator Valve Open
P1549	Wastegate Bypass Regulator Valve Open/Short to Ground
P1550	Charge Pressure Control Difference
P1555	Charge Pressure Upper Limit Exceeded
P1556	Charge Pressure Negative Deviation
P1557	Charge Pressure Positive Deviation
P1558	Throttle Actuator Electrical Malfunction
P1559	Idle Speed Control Throttle Position Adaption Malfunction
P1560	Maximum Engine Speed Exceeded
P1561	Quantity Adjuster Control Difference
P1562	Quantity Adjuster Upper Stop Value
P1563	Quantity Adjuster Lower Stop Value
P1564	Idle Speed Control Throttle Position Low Voltage During Adaption
P1565	Idle Speed Control Throttle Position Lower Limit Not Obtained
P1568	Idle Speed Control Throttle Position Mechanical Malfunction
P1569	Switch For CCS Signal Faulty
P1580	Throttle Actuator B1 Malfunction
P1582	Idle Adaptation At Limit
P1600	Power Supply Terminal No. 15 Low Voltage
P1602	Power Supply Terminal No. 30 Low Voltage
P1603	Internal Control Module Self Check
P1606	Rough Road Spec. Engine Torque ABS-ECU Electrical Malfunction
P1611	MIL Call-Up Circuit/TCM Short to Ground
P1612	Engine Control Module Incorrect Coding
P1613	MIL Call-Up Circuit Open/Short to Positive
P1616	Glow Plug Indicator Lamp Short to Positive
P1617	Glow Plug Indicator Lamp Open/Short to Ground
P1618	Glow Plug Relay Short to Positive
P1619	Glow Plug Relay Open/Short to Ground
P1624	MIL Request Signal Active
P1626	Data Bus Drive Missing Command From M/T
P1630	Accelerator Pedal Position Sensor 1 Signal Too Low
P1631	Accelerator Pedal Position Sensor 1 Signal Too High
P1632	Accelerator Pedal Position Sensor 1/2 Supply Voltage
P1633	Accelerator Pedal Position Sensor 2 Signal Too Low
P1634	Accelerator Pedal Position Sensor 2 Signal Too High
P1639	Accelerator Pedal Position Sensor 1/2 Range Performance
P1640	Internal Control Module (EEPROM) Error
P1648	CAN-Bus System Component Failure
P1649	Data Bus Powertrain Missing Message from Brake Controller
P1676	Drive By Wire MIL Circuit Electrical Malfunction
P1677	Drive By Wire MIL Circuit Short to Positive
P1678	Drive By Wire MIL Circuit Short to Ground
P1679	Drive By Wire MIL Circuit Open Circuit
P1681	Control Module Programming Not Finished
P1686	Control Unit Error, Programming Error
P1690	MIL Malfunction
P1691	MIL Open Circuit
P1692	MIL Short to Ground
P1693	MIL Short to Positive
P1778	Solenoid EV7 Electrical Malfunction
P1780	Engine Intervention Readable
P1851	Data Bus Drive Missing Command from ABS
P1854	Drive Train CAN-Bus Inoperative

Mitsubishi

Code	Description
P1103	Turbocharger Wastegate Actuator
P1104	Turbocharger Wastegate Solenoid
P1105	Fuel Pressure Solenoid
P1300	Ignition Timing Adjustment Circuit
P1400	Manifold Differential Pressure Sensor Circuit
P1500	Alternator FR Terminal Circuit
P1600	Serial Communication Link
P1715	Pulse Generator Assembly
P1750	Solenoid Assembly
P1751	A/T Control Relay
P1791	Engine Coolant Temperature Level Input Circuit
P1795	Throttle Position Input Circuit to TCM